Student Study Guide Part II

CALCULUS

Finney / Thomas

Maurice D. Weir

Naval Postgraduate School

ADDISON-WESLEY PUBLISHING COMPANY
Reading, Massachusetts • Menlo Park, California • New York
Don Mills, Ontario • Wokingham, England • Amsterdam • Bonn
Sydney • Singapore • Tokyo • Madrid • San Juan

Reproduced by Addison-Wesley from camera-ready copy supplied by the author.

ISBN 0-201-53246-8

PREFACE TO THE STUDENT

This study guide has been designed especially for you, the student. It conforms with the textbook, CALCULUS, by Ross L. Finney and George B. Thomas, Jr. It is intended as a self-study workbook to assist you in mastering the basic ideas in calculus. Although this manual was written to conform to the Finney/Thomas CALCULUS, it can be used to accompany any standard calculus textbook and course.

Organization And Learning Objectives

The study manual is organized section by section to correspond with the Finney/Thomas text. For each section we specify its main ideas by stating appropriate learning OBJECTIVES. Each objective states a particular task for you to perform in order to master that objective. Usually the task requires you to solve a certain type of problem related to the discussion in the text; sometimes the task requires you to demonstrate proficiency with certain key terms or concepts. In every case the objective is highly specific and states exactly what you must do.

One or more examples follows each objective and illustrates its requirements. Each example is written in a semi-programmed format; that is, the example is only partially worked out, so you must supply some of the intermediate results yourself. Correct answers to each intermediate result are supplied at the bottom of the page. Thus, each example is broken down into a sequence of steps to guide you through the procedures and techniques associated with its solution. Each example has been carefully selected not to repeat examples or problems in the Finney/Thomas text; thus you retain the full array of the text problems for practice and further learning.

Self-tests

At the end of each chapter there is a SELF-TEST. Each test is followed by complete solutions to all the test problems. The test problems cover the objectives and are similar in scope and difficulty to the examples in this manual and the examples and problems in the Finney/Thomas text. The test should be useful in preparing for class examinations.

How To Use This Study Guide

We recommend that this manual be used in the following way:

1. <u>Read the textbook</u>: Carefully read the section of Finney/Thomas assigned you by your calculus instructor.

2. <u>Study the learning objectives and examples</u>: Read each objective and work through the associated example(s) in the corresponding section of this manual. You should conceal the answers to the examples at the bottom of the page. Work with pencil and scratch paper as you are guided through each solution, writing in the intermediate results in the blanks provided.

3. <u>Check your answers</u>: After all the blanks for a given problem are filled in, compare your answers with the correct answers at the bottom of the page. If you have difficulty or do not fully understand the answers given, review the material in the textbook or consult with your instructor.

4. <u>Do the chapter self-test</u>: After you complete a chapter in Finney/Thomas, review the objectives for that chapter in this manual. Then take the chapter self-test and compare your solutions with those provided. Problems in the self-test sometimes bring together several ideas from the chapter.

Guidance From Your Instructor

We caution you that the learning objectives given in this manual by no means exhaust all the possible objectives that could be written for a careful study of calculus; we have tried to identify the main ones. However, your instructor may have additional requirements. For instance, he or she may want you to be able to prove certain theorems or derive results in the text. We have not stated objectives of this sort. Also, your instructor may consider some objectives far more important than others and not require that you master some objectives at all. So it is imperative that you find out specifically what your instructor considers essential, and study accordingly. This manual should be helpful to you both in identifying the tasks and successfully mastering them. The problems assigned by your instructor should help you discover those concepts and applications of calculus that your instructor wishes to stress.

Maurice D. Weir

TABLE OF CONTENTS

9 Infinite Series 179

9.1 Limits of Sequences of Numbers 179

9.2 Infinite Series 182

9.3 Series without Negative Terms: Comparison and Integral Tests 185

9.4 Series with Nonnegative Terms: Ratio and Root Tests 187

9.5 Alternating Series and Absolute Convergence 187

9.6 Power Series 191

9.7 Taylor Series and Maclaurin Series 193

9.8 Further Calculations with Taylor Series 197

Self-Test 200

Solutions to Self-Test 202

10 Plane Curves and Polar Coordinates 207

10.1 Conic Sections and Quadratic Equations 207

10.2 The Graphs of Quadratic Equations in x and y 210

10.3 Parametric Equations for Plane Curves 212

10.4 The Calculus of Parametric Equations 214

10.5 Polar Coordinates 216

10.6 Graphing in Polar Coordinates 218

10.7 Polar Equations of Conic Sections 220

10.8 Integration in Polar Coordinates 222

Self-Test 225

Solutions to Self-Test 227

11 Vectors and Analytic Geometry in Space 233

11.1 Vectors in the Plane 233

11.2 Cartesian (Rectangular) Coordinates and Vectors in Space 234

11.3 Dot Products 235

11.4 Cross Products 238

11.5 Lines and Planes in Space 240

14.6   Triple Integrals in Cylindrical and Spherical
       Coordinates 309
14.7   Substitutions in Multiple Integrals 313
       Self-Test 315
       Solutions to Self-Test 317

15 Vector Analysis 325

15.1   Line Integrals 325
15.2   Vector Fields, Work, Circulation, and Flux 326
15.3   Green's Theorem in the Plane 330
15.4   Surface Area and Surface Integrals 334
15.5   The Divergence Theorem 338
15.6   Stokes's Theorem 342
15.7   Path Independence, Potential Functions, and Conservative
       Fields 345
       Self-Test 348
       Solutions to Self-Test 350

16 Preview of Differential Equations 357

16.1   Separable First Order Equations 357
16.2   Exact Differential Equations 361
16.3   Linear First Order Equations 362
16.4   Second Order Homogeneous Linear Equations 364
16.5   Second Order Nonhomogeneous Linear Equations 366
16.6   Oscillation 368
16.7   Numerical Methods 370
       Self-Test 371
       Solutions to Self-Test 372

11.6   Surfaces in Space 243

11.7   Cylindrical and Spherical Coordinates 244

        Self-Test 247

        Solutions to Self-Test 248

12 Vector-Valued Functions and Motion in Space 251

12.1   Vector-Valued Functions and Curves in Space, Derivatives and Integrals 251

12.2   Modeling Projectile Motion 254

12.3   Directed Distance and the Unit Tangent Vector $\vec{T}$ 255

12.4   Curvature, Torsion, and the TNB Frame 256

12.5   Planetary Motion and Satellites 261

        Self-Test 263

        Solutions to Self-Test 265

13 Functions of Two or More Variables and Their Derivatives 271

13.1   Functions of Two or More Independent Variables 271

13.2   Limits and Continuity 272

13.3   Partial Derivatives 273

13.4   The Chain Rule 274

13.5   Directional Derivatives and Gradient Vectors 277

13.6   Tangent Planes and Normal Lines 279

13.7   Linearization and Differentials 281

13.8   Maxima, Minima, and Saddle Points 283

13.9   Lagrange Multipliers 286

        Self-Test 288

        Solutions to Self-Test 290

14 Multiple Integrals 297

14.1   Double Integrals 297

14.2   Area, Moments, and Centers of Mass 300

14.3   Double Integrals in Polar Form 303

14.4   Triple Integrals in Rectangular Coordinates. Volumes and Average Values 307

14.5   Masses and Moments in Three Dimensions 308

12. Vector-Valued Functions and Motion in Space 257

13. Functions of Two or More Variables and Their Derivatives 273

14. Multiple Integrals 291

CHAPTER 9 INFINITE SERIES

INTRODUCTION

An ancient Greek paradox, due to the mathematician Zeno, concerned the following problem. Suppose that a man wants to walk a certain distance, say two miles, along a straight line from A to B. First he must pass the half-way point, then the 3/4 point, then the 7/8 point, and so on as illustrated in the following figure.

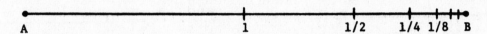

The fractional numbers in the figure indicate the distance in miles remaining to be covered. Therefore, on the assumption that a finite length contains an infinite number of points, the man must pass an infinite number of distance markers along the way. But that seems impossible. The paradox is that the man does get to B, and in a finite amount of time, assuming he walks at some steady pace.

An analysis of the problem is not difficult. The total distance s from A to B is an infinite sum expressible as,

$$s = 1 + \left(\tfrac{1}{2}\right) + \left(\tfrac{1}{2}\right)^2 + \left(\tfrac{1}{2}\right)^3 + \cdots ,$$

and the paradox is dispelled if this infinite sum equals the finite number of 2 miles. That turns out to be exactly the case, as you will see further on in this chapter when it is established that

$$\frac{1}{1 - x} = 1 + x + x^2 + \cdots + x^n + \cdots , \quad \text{if} \quad |x| < 1.$$

For then,

$$2 = \frac{1}{1 - \tfrac{1}{2}} = 1 + \tfrac{1}{2} + \left(\tfrac{1}{2}\right)^2 + \left(\tfrac{1}{2}\right)^3 + \cdots + \left(\tfrac{1}{2}\right)^n + \cdots .$$

9.1 LIMITS OF SEQUENCES OF NUMBERS

OBJECTIVE A: Given a defining rule for the sequence $\{a_n\}$, write the first few items of the sequence.

1. An infinite sequence is a _____ whose domain is the set of _____ greater than or equal to some integer n.

2. The numbers in the range of a sequence are called the _____ of the sequence. The number a_n is called the _____ of the sequence, or the term with _____ n.

---

1. function, integers   2. terms, nth term, index

3. For the sequence whose defining rule is $a_n = 2 + \frac{1}{n}$, the first four terms are

$a_1 = 3$, $a_2 =$ _____, $a_3 =$ _____,   and $a_4 =$ _____.

4. For the sequence whose defining rule is $a_n = \frac{4^n}{n!}$, the first four terms are

$a_1 = 4$, $a_2 =$ _____, $a_3 =$ _____,   and $a_4 =$ _____.

5. For the sequence whose defining rule is $a_n = (-1)^{n+1}\left(\frac{n+1}{n^3}\right)$, the first four terms are

$a_1 = 2$, $a_2 =$ _____, $a_3 =$ _____,   and $a_4 =$ _____.

6. For the sequence defined by the recursion formula $x_{n+1} = \left(\frac{2}{n+1}\right)x_n$, where $x_1 = 2$, the next four terms are

$x_2 =$ _____, $x_3 =$ _____, $x_4 =$ _____,   and $x_5 =$ _____.

OBJECTIVE B : Given a sequence $\{a_n\}$ determine if it converges or diverges. If it converges, use the theorems presented in this article of the text, or l'Hôpital's rule, or Table 9.1, to find the limit.

7. The sequence $\{a_n\}$ converges to the number L if to every _____ ϵ there corresponds an _____ N such that if $n > N$, then _____. If no such limit L exists, we say that $\{a_n\}$ _____.

8. If $0 < b < 1$, then $\{b^n\}$ converges to 0. To see why, consider the inequality

$$|b^n - 0| = b^n < \epsilon.$$

Thus, we seek an integer N such that if $n > N$, then

_____ .

Since the natural logarithm $y = \ln x$ is an increasing function for all x,

$b^n < \epsilon$ is equivalent to $n \ln b < \ln \epsilon$.

Also, because $\ln b$ is a negative number for $0 < b < 1$, this latter inequality is equivalent to

$n > $ _____.

Therefore, we need only choose an integer N satisfying

_____,

and the criterion set forth in Problem 7 for convergence to 0 is satisfied.

---

3. $\frac{5}{2}, \frac{7}{3}, \frac{9}{4}$ 4. $8, \frac{32}{3}, \frac{32}{3}$ 5. $-\frac{3}{8}, \frac{4}{27}, -\frac{5}{64}$ 6. $2, \frac{4}{3}, \frac{2}{3}, \frac{4}{15}$

7. positive number, integer, $|a_n - L| < \epsilon$, diverges   8. $b^n < \epsilon$, $\frac{\ln \epsilon}{\ln b}$, $N > \frac{\ln \epsilon}{\ln b}$

9.   Consider the sequence defined by $a_n = \frac{4^n}{n!}$.   Thus, if $n > 5$,

$$a_n = \frac{4 \cdot 4 \cdot 4 \cdot 4 \cdots 4}{1 \cdot 2 \cdot 3 \cdot 4 \cdots n} = \underline{\hspace{1cm}} \left(\frac{4 \cdot 4 \cdots 4}{5 \cdot 6 \cdots n}\right)$$

$$\leq \frac{32}{3}\left(\frac{4}{5}\right)\underline{\hspace{1cm}} = \left(\frac{32}{3}\right)\left(\frac{5}{4}\right)^4 \left(\frac{4}{5}\right)^n$$

Since $0 \leq a_n$ for all n, the Sandwich Theorem 2 on page 580 of the Finney/Thomas text gives $a_n \to \underline{\hspace{1cm}}$ because $\left(\frac{4}{5}\right)^n \to 0$ from Problem 8.

10.   For the sequence $\left\{\frac{n^3 + 5}{n^2 - 1}\right\}$, $\displaystyle\lim_{n\to\infty} \frac{n^3 + 5}{n^2 - 1} = \lim_{n\to\infty} \frac{n + (5/n^2)}{1 - (1/n^2)} = \underline{\hspace{1cm}}$.
Therefore, the sequence $\underline{\hspace{2cm}}$.

11.   For the sequence defined by $a_n = (-1)^{n+1}\left(\frac{n + 1}{n^3}\right)$,
$0 \leq |a_n| = \underline{\hspace{2cm}} = \frac{1}{n^2} + \underline{\hspace{1.5cm}}$.
Therefore,

$$(-1)^{n+1}\frac{n + 1}{n^3} \to \underline{\hspace{1.5cm}} \quad \text{as} \quad n \to \infty,$$

by the sequence version of the Sandwich Theorem.

12.   Let $a_n = \left(\frac{3n - 1}{5n + 1}\right)^n$.   Then,

$$\frac{3n - 1}{5n + 1} < \frac{3n}{5n + 1} < \underline{\hspace{1.5cm}} \quad \text{implies} \quad 0 \leq \left(\frac{3n - 1}{5n + 1}\right)^n < \underline{\hspace{2cm}}.$$

Therefore $a_n \to \underline{\hspace{1cm}}$ because $\left(\frac{3}{5}\right)^n \to 0$.

13.   Let $a_n = \left(\frac{3n + 1}{5n - 1}\right)^{1/n}$.   Then, applying the natural logarithm to each side, $\ln a_n = \frac{1}{n}(3n + 1) - \underline{\hspace{1.5cm}}$.   By l'Hôpital's rule,

$$\lim_{n\to\infty} \frac{\ln(3n + 1)}{n} = \lim_{n\to\infty} \frac{\underline{\hspace{1.5cm}}}{1} = \underline{\hspace{1.5cm}}, \quad \text{and}$$

$$\lim_{n\to\infty} \frac{\ln(5n - 1)}{n} = \lim_{n\to\infty} \frac{\underline{\hspace{1.5cm}}}{1} = \underline{\hspace{1.5cm}}.$$

Then, $\ln a_n \to 0$ so that taking $f(x) = e^x$ and $L = 0$ in Theorem 3, $a_n = e^{\ln a_n} \to \underline{\hspace{1.5cm}}$.

14.   Consider the sequence defined by $a_n = \left(1 + e^{-n}\right)^n$.   Then,

$$\ln \quad a_n = \underline{\hspace{4cm}} \quad ,$$

and by l'Hôpital's rule,

---

9. $\frac{32}{3}$, n - 4, 0   10. ∞, diverges   11. $\frac{n + 1}{n^3}$, $\frac{1}{n^3}$, 0   12. $\frac{3}{5}$, $\left(\frac{3}{5}\right)^n$, 0

13. $\frac{1}{n}\ln(5n - 1)$, $\frac{3}{3n + 1}$, 0, $\frac{5}{5n - 1}$, 0, $e^0 = 1$

$$\lim_{n \to \infty} \frac{\ln(1 + e^{-n})}{1/n} = \lim_{n \to \infty} \frac{\underline{\hspace{3cm}}}{-1/n^2} \qquad (0/0)$$

$$= \lim_{n \to \infty} \frac{n^2}{e^n + 1} \qquad (\infty/\infty)$$

$$= \lim_{n \to \infty} \underline{\hspace{3cm}} \qquad (\text{still } \infty/\infty)$$

$$= \lim_{n \to \infty} \frac{2}{e^n} = \underline{\hspace{2cm}}.$$

Therefore,

$$a_n = e^{\ln a_n} \to \underline{\hspace{2cm}}.$$

15. Let $a_n = \sqrt[n]{n^3}$. Then, $a_n = \left(n^3\right)^{1/n} = n^{\underline{\hspace{0.6cm}}} = \left(\sqrt[n]{n}\right)^{\underline{\hspace{0.6cm}}}$.

Now, $\sqrt[n]{n} \to \underline{\hspace{2cm}}$ by limit two in Table 9.1 in the text, page 582. Thus, if $f(x) = x^3$, then

$$a_n = f(\sqrt[n]{n}) \to \underline{\hspace{2cm}}.$$

9.2   INFINITE SERIES

16. If $\{a_n\}$ is a sequence, and

$$s_n = a_1 + a_2 + \ldots + a_n,$$

then the sequence $\{s_n\}$ is called an $\underline{\hspace{4cm}}$.

17. The number s_n is called the $\underline{\hspace{4cm}}$ of the series.

18. Instead of $\{s_n\}$ we usually use the notation $\underline{\hspace{2cm}}$ for the series.

19. The series $\sum_{n=1}^{\infty} a_n$ is said to converge if the sequence

$\underline{\hspace{2cm}}$ converges to a finite limit L. In that case we write $\underline{\hspace{3cm}}$ or $a_1 + a_2 + \ldots + a_n + \ldots = L$. If no such limit exists, the series is said to $\underline{\hspace{3cm}}$.

OBJECTIVE A : For a given geometric series $\sum_{n=1}^{\infty} ar^{n-1}$, determine if the series converges or diverges. If it does converge, then compute the sum of the series. The indexing of the series may be changed for a given problem.

20. Consider the series $\sum_{n=1}^{\infty} \frac{3}{5^{n-1}}$. This is a geometric series with

$a = \underline{\hspace{2cm}}$ and $r = \underline{\hspace{2cm}}$. Since $|r| < 1$, the geometric

---

14. $n \ln(1 + e^{-n})$, $\left(\frac{1}{1 + e^{-n}}\right)(-e^{-n})$, $\frac{2n}{e^n}$, 0, $e^0 = 1$ 15. $\frac{3}{n}$, 3, 1, $1^3 = 1$ 16. infinite series

17. nth partial sum   18. $\sum_{n=1}^{\infty} a_n$ 19. $\{s_n\}$, $\sum_{n=1}^{\infty} a_n = L$, diverge

series _____, and its sum is given by

$$\sum_{n=1}^{\infty} \frac{3}{5^{n-1}} = \underline{\hspace{2cm}} = \underline{\hspace{2cm}}.$$

21. The series $\sum_{n=3}^{\infty} (-1)^{n-1} \frac{4}{3^{n-1}}$ is a geometric series with

a = _____ and r = _____. Since $|r| < 1$, the geometric series _____. However, the index begins with n = 3 instead of n = 1. Now,

$$\sum_{n=3}^{\infty} (-1)^{n-1} \frac{4}{3^{n-1}} = \sum_{n=1}^{\infty} (-1)^{n-1} \frac{4}{3^{n-1}} - \left(\underline{\hspace{2cm}} \right)$$

$$= \underline{\hspace{2cm}} - \frac{8}{3} = \underline{\hspace{2cm}}.$$

22. The series $\sum_{n=3}^{\infty} \frac{2^n}{7}$ is a geometric series with a = _____ and

r = _____. Since _____ the series diverges.

23. The repeating decimal 0.15 15 15 ... is a geometric series in disguise. It can be written as

$$0.15\ 15\ 15\ ... = \frac{15}{100} + \frac{15}{\underline{\hspace{1cm}}} + \frac{15}{\underline{\hspace{1cm}}} + ...$$

$$= \sum_{n=1}^{\infty} \underline{\hspace{2cm}}$$

$$= \frac{1}{100} \sum_{n=1}^{\infty} \underline{\hspace{2cm}}$$

$$= \frac{1}{100} \left(\underline{\hspace{2cm}} \right) = \frac{15}{\underline{\hspace{1cm}}} = \frac{15}{99}.$$

24. If you write in the first four terms of the following geometric series you have

$$\sum_{n=0}^{\infty} ar^n = \underline{\hspace{1.5cm}} + \underline{\hspace{1.5cm}} + \underline{\hspace{1.5cm}} + \underline{\hspace{1.5cm}} + ...$$

Thus, $\sum_{n=0}^{\infty} ar^n$ is simply another way of writing the geometric

series $\sum_{n=1}^{\infty} ar^{n-1}$. Therefore, if $|r| < 1$ we conclude that

$$\sum_{n=0}^{\infty} ar^n = \underline{\hspace{2cm}}.$$

---

20. 3, $\frac{1}{5}$, converges, $\frac{3}{1 - \frac{1}{5}}$, $\frac{15}{4}$

21. 4, $-\frac{1}{3}$, converges, $4 - \frac{4}{3}$, $\frac{4}{1 + \frac{1}{3}}$, $\frac{1}{3}$

22. $\frac{1}{7}$, 2, $|r| > 1$

23. 100^2, 100^3, $\frac{15}{100^n}$, $\frac{15}{100^{n-1}}$, $\frac{15}{1 - \frac{1}{100}}$, 100 - 1

24. a, ar, ar^2, ar^3, $\frac{a}{1 - r}$

25. Sometimes the terms of a given series are a sum or difference of terms, each of which beongs to a geometric series. For example,

$$\sum_{n=0}^{\infty} \left(\frac{7}{3^n} - \frac{1}{2^n}\right) = \sum_{n=0}^{\infty} \frac{7}{3^n} - \sum_{n=0}^{\infty} \frac{1}{2^n}$$

$$= \underline{\hspace{1.5cm}} - \underline{\hspace{1.5cm}} = \underline{\hspace{1.5cm}} - \frac{4}{2} = \underline{\hspace{1.5cm}} .$$

$\boxed{\text{OBJECTIVE B}}$: Use the nth-term test for divergence to test a given series $\sum_{n=1}^{\infty} a_n$ for divergence.

26. For the series $\sum_{n=1}^{\infty} \frac{n^n}{n!}$, we have for every index n,

$$a_n = \frac{n^n}{n!} = \frac{n \cdot n \cdot n \cdots n}{1 \cdot 2 \cdot 3 \cdots n} = \left(\frac{n}{1}\right) \left(\frac{n}{2}\right) \left(\underline{\hspace{1cm}}\right) \cdots \left(\frac{n}{n}\right) > \underline{\hspace{1.5cm}} .$$

Therefore, $\lim_{n \to \infty} a_n \neq 0$. We conclude that the series $\underline{\hspace{1.5cm}}$.

27. Consider the series $\sum_{n=1}^{\infty} (-1)^{n+1} \left(1 + \frac{1}{3^n}\right)$. For large values of the index n, the absolute value of the nth term,

$$|a_n| = 1 + 3^{-n}$$

is close to $\underline{\hspace{1.5cm}}$. Therefore, the limit

$$\lim_{n \to \infty} (-1)^{n+1} \left(1 + \frac{1}{3^n}\right) \underline{\hspace{1.5cm}} \text{ exist. We conclude that the}$$

series $\underline{\hspace{1.5cm}}$.

28. For the series $\sum_{n=1}^{\infty} \left(\frac{n + 2}{n}\right)^n$, we have

$$\lim_{n \to \infty} \left(\frac{n + 2}{n}\right)^n = \lim_{n \to \infty} \left(1 + \frac{2}{n}\right)^n = \underline{\hspace{1.5cm}} .$$ Thus, the series

$\underline{\hspace{2cm}}$ because the limit of the nth term is not zero.

29. For the series $\sum_{n=1}^{\infty} \frac{2n^2 + 1}{n^3 - 1}$ the limit of the nth-term is

$$\lim_{n \to \infty} \left(\frac{2n^2 + 1}{n^3 - 1}\right) = \lim_{n \to \infty} \left(\frac{\frac{2}{n} + \frac{1}{n^3}}{\underline{\hspace{1cm}}}\right) = \frac{0 + 0}{\underline{\hspace{1cm}}} = 0 .$$

Since the limit is equal to zero, the nth-term test gives no information concerning convergence or divergence.

---

25. $\frac{7}{1 - \frac{1}{3}}$, $\frac{1}{1 - \frac{1}{2}}$, $\frac{21}{2}$, $\frac{17}{2}$ 26. $\frac{n}{3}$, 1, diverges   27. 1, does not, diverges

28. e^2, diverges   29. $1 - \frac{1}{n^3}$, $1 + 0$

OBJECTIVE C : Use partial fractions to find the sum of a series, if appropriate.

30. Consider the series $\displaystyle\sum_{n=1}^{\infty} \frac{1}{(2n-1)(2n+1)}$. This is not a geometric series. However, we can use partial fractions to re-write the kth term:

$$\frac{1}{(2k-1)(2k+1)} = \frac{1}{2}\left[\underline{\hspace{2cm}} - \frac{1}{2k+1} \right].$$

This permits us to write the partial sum

$$\sum_{n=1}^{k} \frac{1}{(2n-1)(2n+1)} = \frac{1}{1 \cdot 3} + \frac{1}{3 \cdot 5} + \cdots + \frac{1}{(2k-1)(2k+1)}$$

as

$$s_k = \frac{1}{2}\left(\frac{1}{1} - \frac{1}{3}\right) + \frac{1}{2}\left(\underline{\hspace{1.5cm}}\right) + \frac{1}{2}\left(\underline{\hspace{1.5cm}}\right)$$
$$+ \cdots + \frac{1}{2}\left(\frac{1}{2k-1} - \frac{1}{2k+1}\right).$$

By removing parentheses on the right, and combining terms, we find that

$$s_k = \underline{\hspace{4cm}}.$$

Therefore, $s_k \to$ _____ and the series _____ converge. Hence,

$$\sum_{n=1}^{\infty} \frac{1}{(2n-1)(2n+1)} = \underline{\hspace{1.5cm}}.$$

9.3 SERIES WITHOUT NEGATIVE TERMS: COMPARISON AND INTEGRAL TESTS

OBJECTIVE A : Know the Nondecreasing Sequence Theorem and how it applies to an infinite series of nonnegative terms.

31. A nondecreasing sequence $\{s_n\}$ is a sequence with the property that _____ for every n.

32. A sequence $\{s_n\}$ is said to be _____ from above if there is a finite constant M such that $s_n \leq M$ for every n.

33. If $\{s_n\}$ is a nondecreasing sequence that is bounded from above, then it _____.

34. If a nondecreasing sequence $\{s_n\}$ fails to be bounded, then it _____.

35. If Σa_n is a series of nonnegative terms, then its sequence $\{s_n\}$ of partial sums is _____.

---

30. $\frac{1}{2k-1}$, $\frac{1}{3} - \frac{1}{5}$, $\frac{1}{5} - \frac{1}{7}$, $\frac{1}{2}\left(1 - \frac{1}{2k+1}\right)$, $\frac{1}{2}$, does, $\frac{1}{2}$ 31. $s_n \leq s_{n+1}$ 32. bounded

33. converges   34. diverges to plus infinity   35. nondecreasing

36.   If Σa_n is a series of nonnegative terms, then it converges if and only if its sequence $\{s_n\}$ of partial sums is _____ _____.

[OBJECTIVE B]:   Use the comparison and integral tests for convergence and divergence to determine whether a given series with no negative terms converges or diverges.

37. $\displaystyle\sum_{n=1}^{\infty} \frac{n + 5}{n^2 - 3n + 5}$

For every index n, $5n > -3n + 5$ because n is a positive integer. Then, $n^2 + 5n >$ _____, and since $n^2 - 3n + 5$ is positive for every index it follows that

$$\frac{n + 2}{n^2 - 3n + 5} > \underline{\hspace{2cm}}.$$

We conclude that the given series _____ by comparison with the series $\sum \frac{1}{n}$.

38. $\displaystyle\sum_{n=2}^{\infty} \frac{1}{n(\ln n)^2}$

We apply the integral test $\displaystyle\int_{2}^{\infty} \frac{dx}{x(\ln x)^2} = \lim_{b \to \infty} \underline{\hspace{2cm}} \Big]_{b}^{2} = \underline{\hspace{1.5cm}}.$

Therefore, the integral _____ and hence the series _____.

39. $\displaystyle\sum_{n=1}^{\infty} \frac{(\ln n)^2}{n^3}$

The following argument shows that $\ln n < \sqrt{n}$ for every index n. First, define the function $g(x) = \sqrt{x} - \ln x$. Now $g'(x) =$ _____ is nonnegative if $x \geq$ _____, and it follows that g is an increasing function of x for $x \geq 4$. Also, $g(4) = 2 - \ln 4 \approx 0.613$ is positive. Therefore, $g(x) > 0$ for $x \geq 4$. A simple verification using tables, or a calculator, shows that $g(1)$, $g(2)$, and $g(3)$ are positive. Hence, we have established that $\ln n < \sqrt{n}$ for every positive integer n. Using this fact,

$$\frac{(\ln n)^2}{n^3} < \frac{(\sqrt{n})^2}{n^3} = \underline{\hspace{2cm}}.$$

We conclude that the series $\displaystyle\sum_{n=1}^{\infty} \frac{(\ln n)^2}{n^3}$ _____ by the comparison test.

---

36.   bounded from above

37. $n^2 - 3n + 5$, $\frac{1}{n}$, diverges

38. $\frac{1}{\ln x}$, $\frac{1}{\ln 2}$, converges, converges

39. $\frac{1}{2\sqrt{x}} - \frac{1}{x}$, 4, $\frac{1}{n^2}$, converges

9.4 SERIES WITH NONNEGATIVE TERMS: RATIO AND ROOT TESTS

[OBJECTIVE]: Given a series with nonnegative terms, investigate convergence or divergence using the ratio and root tests.

40. $\sum_{n=1}^{\infty} \frac{n!}{3^n}$

We try the ratio test. Thus, $\frac{a_{n+1}}{a_n} = \frac{(n+1)!/3^{n+1}}{n!/3^n} = $ _____ .

Hence, $\rho = \lim_{n \to \infty} \frac{a_{n+1}}{a_n} = $ _____ , and the series _____ .

41. $\sum_{n=1}^{\infty} (\sqrt[n]{n} - 1)^n$

We try the root test. Thus, for $a_n = (\sqrt[n]{n} - 1)^n$,

$$\sqrt[n]{a_n} = \underline{\hspace{2cm}} \to \underline{\hspace{2cm}}.$$

Because $\rho = $ _____ we conclude that the series _____ according to the root test.

42. $\sum_{n=1}^{\infty} \frac{n \ln n}{(n+1)^2}$

We try the ratio test. Thus,

$$\frac{a_{n+1}}{a_n} = \frac{(n+1) \ln (n+1)}{(n+2)^2} \cdot \underline{\hspace{2cm}}.$$

Hence,

$$\rho = \lim_{n \to \infty} \frac{a_{n+1}}{a_n} = \lim_{n \to \infty} \left[\frac{(n+1)^3}{n(n+1)^2} \cdot \underline{\hspace{1.5cm}} \right] = \underline{\hspace{1cm}} \cdot 1 = \underline{\hspace{1cm}}.$$

In this case the root test says the series _____ _____ .

9.5 ALTERNATING SERIES AND ABSOLUTE CONVERGENCE

[OBJECTIVE A]: Use the Alternating Series Theorem to investigate the convergence of an alternating series.

43. $\sum_{n=1}^{\infty} (-1)^{n+1} \frac{n}{n^2 + 1}$

First we see that $a_n = \frac{n}{n^2 + 1}$ is positive for every n. Also, $\lim_{n \to \infty} a_n = $ _____ . Next, we compare a_{n+1} with a_n for

---

40. $\frac{1}{3} \cdot (n+1)$, $+\infty$, diverges   41. $\sqrt[n]{n} - 1$, 0, 0, converges

42. $\frac{(n+1)^2}{n \ln n}$, $\frac{\ln (n+1)}{\ln n}$, 1, 1, may converge or may diverge

arbitrary n. Now, $\dfrac{n+1}{(n+1)^2+1} \le \dfrac{n}{n^2+1}$ if and only if

$(n+1)(n^2+1) \le n[(n+1)^2+1]$. This last inequality is equivalent to $n^3 + n^2 + n + 1 \le \underline{\hspace{2cm}}$ or, $1 \le n^2 + n$ which is true. We conclude that the alternating series

$\underline{\hspace{2cm}}$.

44. $\displaystyle\sum_{n=1}^{\infty} (-1)^{n+1} \dfrac{1}{n^{1+1/n}}$

It is clear that $a_n = \dfrac{1}{n^{1+1/n}} = \dfrac{1}{n \cdot \sqrt[n]{n}}$ is positive for every

n. Also, $\displaystyle\lim_{n\to\infty} a_n = \lim_{n\to\infty} \dfrac{1}{n} \cdot \lim_{n\to\infty} \dfrac{1}{\sqrt[n]{n}} = \underline{\hspace{2cm}}$. We would

like to show that $\{a_n\}$ is a $\underline{\hspace{2cm}}$ sequence. One way is to replace n by the continuous variable x and show that the resultant function

$$y = f(x) = x^{1+1/x}, \quad x > 0,$$

which is the reciprocal of a_n for x = n, is an <u>increasing</u> function of x for every x. If we take the logarithm of both sides of this last equation, and differentiate implicitly with respect to x, we obtain

$$\ln y = \left(1 + \tfrac{1}{x}\right) \ln x, \quad \text{and} \quad \frac{y'}{y} = \frac{1}{x^2}(\underline{\hspace{2cm}}) + \frac{1}{x}.$$

Simplifying algebraically, $y' = \dfrac{y}{x^2}(1 + \underline{\hspace{1.5cm}} - \ln x)$.

Thus, since $x > \ln x$ for all $x > 0$, we find that y' is positive, and $y = f(x)$ is increasing for all x (so the reciprocal is $\underline{\hspace{2cm}}$). Therefore we have established that

$$\frac{1}{(n+1) \cdot \sqrt[n+1]{n+1}} < \frac{1}{n \cdot \sqrt[n]{n}} \quad \text{for every index} \quad n.$$

We conclude that the alternating series $\underline{\hspace{3cm}}$.

OBJECTIVE B: Use the Alternating Series Estimation Theorem to estimate the magnitude of the error if the first k terms, for some specified number k, are used to approximate a given alternating series.

45.   It can be shown, with a little work, that the alternating harmonic series

$$\sum_{n=1}^{\infty} \frac{(-1)^{n+1}}{n}$$

converges to ln 2. If we wish to approximate ln 2 correct to four decimal places using this series, the alternating series error estimation gives

---

_____ $< 0.5 \times 10^{-5}$,   or   n > _____.

Therefore, we would need to sum the first 200,000 terms of the alternating harmonic series to <u>ensure</u> four decimal place accuracy in approximating ln 2. This does not mean that fewer terms would <u>not</u> provide that accuracy. A more efficient approximation for ln 2, accurate to four decimal places, uses Simpson's rule with n = 6 to estimate

$$\int_1^2 \frac{dx}{x}.$$

OBJECTIVE C : Given an infinite series, use the tests studied in this chapter of the text to determine if the series is absolutely convergent, conditionally convergent, or divergent.

46. A series $\sum_{n=1}^{\infty} a_n$ is said to converge absolutely if

_____.

47. True or False:
(a) If a series converges absolutely, then it converges.
(b) If a series converges, then it converges absolutely.

48. If a series $\sum a_n$ converges, but the series of absolute values $\sum |a_n|$ diverges, we say that the original series $\sum a_n$ is

_____.

49. $\sum_{n=1}^{\infty} \frac{(-1)^{n+1} n \ln n}{3^n}$

The absolute value of the nth term of the series is $|a_n| = \frac{n \ln n}{3^n}$.   Applying the ratio test,

$\frac{|a_{n+1}|}{|a_n|} = \frac{(n+1) \ln (n+1)/3^{n+1}}{n \ln n/3^n} = $ _____.   By l'Hôpital's

rule, $\lim_{n\to\infty} \frac{\ln (n+1)}{\ln n} = \lim_{n\to\infty}$ _____ = _____.

Therefore, $\lim_{n\to\infty} \frac{|a_{n+1}|}{|a_n|} = $ _____, so we conclude that

$\sum_{n=1}^{\infty} \frac{(-1)^{n+1} n \ln n}{3^n}$ _____ converge absolutely.

---

45. $\frac{1}{n}$, 2×10^5 46. $\sum_{n=1}^{\infty} |a_n|$ converges   47. (a) True (b) False

48. conditionally convergent   49. $\frac{1}{3}\left(\frac{n+1}{n}\right) \frac{\ln (n+1)}{\ln n}$, $\frac{1/(n+1)}{1/n}$, 1, $\frac{1}{3}$, does

50. $\displaystyle\sum_{n=1}^{\infty} \frac{2n - n^2}{n^3}$

The nth term of the series is $a_n = \dfrac{2n - n^2}{n^3} = \dfrac{n(2 - n)}{n^3}$ which is

negative if $n >$ _____ . Thus, $-a_n = \dfrac{n - 2}{n^2}$ is positive for

$n > 2$. Now, $\dfrac{n - 2}{n^2} > \dfrac{1}{2n}$ whenever $n >$ _____ . Since the

series $\displaystyle\sum_{n=1}^{\infty} \frac{1}{2n}$ _____ , we conclude that the series

$\displaystyle\sum_{n=1}^{\infty} |a_n| = \sum_{n=1}^{\infty} \frac{n - 2}{n^2}$ _____ by the comparison test.

Therefore, the original series _____ converge absolutely.

51. In Problem 44 above, the series $\displaystyle\sum_{n=1}^{\infty} (-1)^{n+1} \frac{1}{n^{1+ 1/n}}$ was shown to

be convergent. We want to know if the series converges
absolutely. Using the same technique as in Problem 44, it is
easy to establish that

$$^{n+1}\sqrt{n + 1} < \; ^n\sqrt{n} \quad \text{if} \quad n \geq 3:$$

we define the function $y = x^{1/n}$, $x \geq 3$, and show that y' is
always negative; whence we conclude that y is a _____
function of x . In particular, $^n\sqrt{n} < \; ^3\sqrt{3} \approx 1.44 < \dfrac{3}{2}$ if

$n \geq$ _____ . It follows that

$$\frac{1}{n \cdot \; ^n\sqrt{n}} > \underline{\hspace{2cm}} \quad \text{if} \quad n \geq 4.$$

Since the harmonic series $\sum \frac{1}{n}$ diverges, we find that the

series $\displaystyle\sum_{n=1}^{\infty} \frac{1}{n^{1+ 1/n}}$ _____ by the comparison test.

Therefore, the original series $\displaystyle\sum_{n=1}^{\infty} (-1)^{n+1} \frac{1}{n^{1+ 1/n}}$ is

_____ .

52. $\displaystyle\sum_{n=1}^{\infty} (-1)^{n+1} \left(\frac{n + 1}{n}\right)^n$

In this case, $a_n = \left(\dfrac{n + 1}{n}\right)^n = \left(1 + \dfrac{1}{n}\right)^n \to$ _____ . Therefore

the series _____ .

53. $\displaystyle\sum_{n=2}^{\infty} \frac{\cos n\pi}{n\sqrt{\ln n}}$

Since $\cos n\pi$ is 1 when n is even, and -1 when n is
odd, the series is alternating in sign. Let us see if the
series converges absolutely. Now,

---

50. 2, 4, diverges, diverges, does not

51. decreasing, 4, $\frac{2}{3n}$, diverges, conditionally convergent   52. e, diverges

$$|a_n| = \left|\frac{\cos n\pi}{n\sqrt{\ln n}}\right| = \underline{\hspace{2cm}}.$$ Applying the integral test,

$$\int_1^\infty \frac{dx}{x\sqrt{\ln x}} = \lim_{b\to\infty} \underline{\hspace{2cm}} \Big]_1^b = \underline{\hspace{2cm}}.$$

Therefore, the series $\sum |a_n|$ _____. To see if the original series converges we check the three conditions of the Alternating Series Theorem (remember that the numerator $\cos n\pi$ simply determines the <u>sign</u> of the nth term of the series): $\frac{1}{n\sqrt{\ln n}}$ is positive for all n, and converges to _____.

It is clear that $\frac{1}{(n+1)\sqrt{\ln (n+1)}} < \frac{1}{n\sqrt{\ln n}}$ because $y = \ln x$ is an _____ function of x. Therefore, the series $\sum_{n=1}^\infty \frac{\cos n\pi}{n\sqrt{\ln n}}$ is _____.

9.6   POWER SERIES

54. A series of the form

$$\sum_{n=0}^\infty c_n x^n = c_0 + c_1 x + c_2 x^2 + c_3 x^3 + \ldots + c_n x^n + \ldots$$

is called a _____.

55. For the power series

$$\sum_{n=0}^\infty c_n(x-a)^n = c_0 + c_1(x-a) + c_2(x-a)^2 + \ldots + c_n(x-a)^n + \ldots$$

the constant a is called the _____. The coefficients $c_0, c_1, c_2, \ldots, c_n, \ldots$ are all _____.

OBJECTIVE A : Given a power series $\sum_{n=0}^\infty a_n x^n$, find its interval of convergence. If the interval is finite, determine whether the series converges at each endpoint.

56. $\sum_{n=1}^\infty \frac{1}{\sqrt{n}\,3^n} x^n$

We apply the ratio test to the series of absolute values, and find

$$\rho = \lim_{n\to\infty} \left|\frac{x^{n+1}}{\sqrt{n+1}\,3^{n+1}} \cdot \underline{\hspace{2cm}}\right| = \lim_{n\to\infty} \frac{\sqrt{n}}{\underline{\hspace{1cm}}} |x| = \underline{\hspace{2cm}}.$$

---

53. $\frac{1}{n\sqrt{\ln n}}$, $2\sqrt{\ln x}$, $+\infty$, diverges, 0, increasing, conditionally convergent

54. power series   55. center, constants

Therefore the original series converges absolutely if $|x| < \underline{\hspace{1.5cm}}$ and diverges if $\underline{\hspace{1.5cm}}$. When $x = 3$, the series becomes

$$\sum_{n=1}^{\infty} \underline{\hspace{2cm}}, \quad \text{the p-series with } p = \underline{\hspace{1.5cm}};$$

this series $\underline{\hspace{1.5cm}}$. When $x = -3$, the series becomes

$$\sum_{n=1}^{\infty} \underline{\hspace{2cm}},$$

and this series $\underline{\hspace{1.5cm}}$, by the Alternating Series Theorem. Therefore, the interval of convergence of the original power series is $\underline{\hspace{2cm}}$.

57. $\displaystyle\sum_{n=1}^{\infty} \frac{2^n}{n(3^{n+2})} x^{n+1}$

The power series converges for $x = 0$. For $x \neq 0$, we apply the root test to the series of absolute values, and find

$$\rho = \lim_{n\to\infty} \sqrt[n]{\frac{2^n \, |x|^n \, |x|}{n \cdot 3^n \cdot 3^2}} = \lim_{n\to\infty} \underline{\hspace{4cm}}$$

$$= \frac{2|x| \cdot 1}{\underline{\hspace{1.5cm}}} < 1, \quad \text{if} \quad |x| < \underline{\hspace{1.5cm}}.$$

Therefore, the original series converges absolutely if $|x| < \frac{3}{2}$ and diverges if $|x| > \frac{3}{2}$. When $x = \frac{3}{2}$, the series becomes

$$\sum_{n=1}^{\infty} \frac{2^n}{n(3^{n+2})} \left(\frac{3}{2}\right)^{n+1} = \sum_{n=1}^{\infty} \underline{\hspace{2cm}},$$

and this series $\underline{\hspace{1.5cm}}$. When $x = -\frac{3}{2}$, the series becomes

$$\sum_{n=1}^{\infty} \frac{(-1)^{n+1}}{6n},$$

and this series $\underline{\hspace{1.5cm}}$, by the Alternating Series Theorem. Therefore, the interval of convergence of the original power series is $\underline{\hspace{2cm}}$.

58. $\displaystyle\sum_{n=1}^{\infty} [\sin(5n)](x - \pi)^n$

For every value of x, $|[\sin(5n)](x - \pi)^n| \leq |x - \pi|^n$. The geometric series $\displaystyle\sum_{n=1}^{\infty} |x - \pi|^n$ converges if $\underline{\hspace{2cm}}$ and diverges if $\underline{\hspace{2cm}}$. Therefore, by the comparison test, the original series converges absolutely if $\underline{\hspace{2cm}}$.

---

56. $\dfrac{\sqrt{n}\, 3^n}{x^n}$, $3\sqrt{n+1}$, $\frac{1}{3}|x|$, 3, $|x| > 3$, $\frac{1}{\sqrt{n}}$, $\frac{1}{2}$, diverges, $\dfrac{(-1)^n}{\sqrt{n}}$, converges, $-3 \leq x < 3$

57. $\dfrac{2\,|x| \cdot \sqrt[n]{|x|}}{n\sqrt{n} \cdot 3 \cdot \sqrt[n]{9}}$, $1 \cdot 3 \cdot 1$, $\frac{3}{2}$, $\frac{1}{6n}$, diverges, converges, $-\frac{3}{2} \leq x < \frac{3}{2}$

Suppose $|x - \pi| = 1$.   Then the series becomes,

$$\sum_{n=1}^{\infty} \sin (5n) \quad \text{or} \quad \sum_{n=1}^{\infty} \underline{\hspace{3cm}}.$$

However, $\lim_{n \to \infty} \sin (5n)$ fails to exist, so neither of these series can converge. We conclude that the interval of convergence of the original series is $\underline{\hspace{5cm}}$.

OBJECTIVE B: Given a power series $f(x) = \sum a_n x^n$,   find the power series for $f'(x)$.

59.   In Example 4, on page 620 of the text, it is given that

$$\frac{1}{1 + x^2} = 1 - x^2 + x^4 - x^6 + \cdots , \quad \text{for} -1 < x < 1.$$

Therefore, using the term-by-term differentiation theorem,

$$\frac{-2x}{\left(1 + x^2\right)^2} = \underline{\hspace{4cm}}, \quad \text{for} \quad \underline{\hspace{3cm}}.$$

9.7   TAYLOR SERIES AND MACLAURIN SERIES

OBJECTIVE A: Find the Taylor Series at $x = a$,   or the Maclaurin series, for a given function $y = f(x)$. Assume that $x = a$ is specified and that f has finite derivatives of all orders at $x = a$.

60.   If $y = f(x)$ has finite derivatives of all orders at $x = a$, the particular power series

$$f(a) + f'(a)(x - a) + \frac{f''(a)}{2!}(x - a)^2 + \cdots + \frac{f^{(n)}(a)}{n!}(x - a)^n + \cdots$$

is called the $\underline{\hspace{6cm}}$.
If $a = 0$,   the series is known as the $\underline{\hspace{3cm}}$ for f. The Taylor series for a function may or may not converge to the function.

61.   If $y = f(x)$ has finite derivatives of order up to and including n,   then the polynomial

$$P_n(x) = f(a) + f'(a)(x - a) + \frac{f''(a)}{2!}(x - a)^2 + \cdots + \frac{f^{(n)}(a)}{n!}(x-a)^n$$

is called the $\underline{\hspace{6cm}}$
$\underline{\hspace{2cm}}$. The graph of this polynomial passes through the point $\underline{\hspace{1.5cm}}$,   and its first n derivatives match the first

---

58. $|x - \pi| < 1$, $|x - \pi| \geq 1$, $|x - \pi| < 1$, $(-1)^n \sin (5n)$, $\pi - 1 < x < \pi + 1$

59. $-2x + 4x^3 - 6x^5 + \cdots$, $-1 < x < 1$

60.   Taylor series generated by f at $x = a$, Maclaurin series

n derivatives of _____ at _____. Each nonnegative
integer n corresponds to a Taylor polynomial for f at
x = a, provided the first n derivatives of f exist at
x = a.

62.   Let us find the Maclaurin series for the function
$f(x) = x^5 + 4x^4 + 3x^3 + 2x + 1$. We need to find the derivatives
of f of all orders, and evalute them at x = 0:

$f'(x)$ = _____ , $f'(0)$ = 2

$f^{(2)}(x)$ = _____ , $f^{(2)}(0)$ = _____

$f^{(3)}(x)$ = _____ , $f^{(3)}(0)$ = _____

$f^{(4)}(x)$ = 120x + 96   , $f^{(4)}(0)$ = 96

$f^{(5)}(x)$ = _____   , $f^{(5)}(0)$ = _____

$f^{(6)}(x)$ = 0   , $f^{(6)}(0)$ = 0

In general, $f^{(k)}(0)$ = _____ if k ≥ 6. Thus, the Maclaurin
series is

_____ ,

which simplifies to $1 + 2x + 3x^3 + 4x^4 + x^5$. Therefore, the
Maclaurin series for a polynomial expressed in powers of x is
the polynomial itself.

63.   Suppose we want to express the polynomial in Problem 62 in
powers of (x + 1) instead of powers of x. We find the
Taylor series of f at x = _____. From our previous
calculations of the derivatives, we find that

$f(-1) = -1$, $f'(-1) = 0$, $f^{(2)}(-1)$ = _____, $f^{(3)}(-1)$ = _____,
$f^{(4)}(-1)$ = _____, $f^{(5)}(-1)$ = _____, and $f^{(k)}(-1)$ = _____
if k ≥ 6. Thus the Taylor series of f at x = -1 is

_____ ,

which simplifies to

$-1 + 5(x + 1)^2 - 3(x + 1)^3 - (x + 1)^4 + (x + 1)^5$.

64.   Let us find the Taylor polynomials $P_3(x)$ and $P_4(x)$ for the
function $f(x) = a^x$, a > 0, at x = 1. To do this we complete
the following table:

_____

61. Taylor polynomial of order n generated by f at x = a, (a, f(a)), y = f(x), x = a

62. $5x^4 + 16x^3 + 9x^2 + 2$, $20x^3 + 48x^2 + 18x$, 0, $60x^2 + 96x + 18$, 18, 120, 120, 0,

   $1 + 2x + 0x^2 + \frac{18}{3!}x^3 + \frac{96}{4!}x^4 + \frac{120}{5!}x^5$

63. -1, 10, -18, -24, 120, 0, $-1 + 0(x + 1) + \frac{10}{2!}(x + 1)^2 - \frac{18}{3!}(x + 1)^3 - \frac{24}{4!}(x + 1)^4 + \frac{120}{5!}(x + 1)^5$

n	$f^{(n)}(x)$	$f^{(n)}(1)$
0	a^x	a
1	$a^x \ln a$	$a \ln a$
2	_____	_____
3	_____	_____
4	_____	_____

Then,

$$P_3(x) = a + a(\ln a)(x - 1) + \frac{a(\ln a)^2}{2!}(x - 1)^2 + \underline{\hspace{3cm}},$$

$$P_4(x) = \underline{\hspace{8cm}}.$$

65. For the function $f(x) = a^x$ in Problem 64, the Taylor series at $x = 1$ is

$$\sum_{n=0}^{\infty} \underline{\hspace{4cm}}.$$

[OBJECTIVE B]: Know the statement of Taylor's Theorem and a formula for the remainder of order n.

66. The statement of Taylor's Theorem in the text gives the remainder term as

$$R_n(x) = \underline{\hspace{5cm}},$$

where the number c lies between $\underline{\hspace{4cm}}$. This remainder term measures the error in the approximation of $y = f(x)$ by the nth-degree Taylor polynomial at $\underline{\hspace{2cm}}$. Thus, the Taylor series expansion for $f(x)$ will converge to $f(x)$ provided that

$$\underline{\hspace{5cm}}.$$

67. This remainder form is very useful because often we can bound the derivative $f^{(n+1)}(c)$ by some constant M: $|f^{(n+1)}(c)| \leq M$. This ensures that $R_n(x)$ converges to $\underline{\hspace{1cm}}$ as $n \to \infty$.

---

64. For $n = k$, $f^{(k)}(1) = a(\ln a)^k$; $\frac{a(\ln a)^3}{3!}(x - 1)^3$,

   $a + a(\ln a)(x - 1) + \frac{a(\ln a)^2}{2!}(x - 1)^2 + \frac{a(\ln a)^3}{3!}(x - 1)^3 + \frac{a(\ln a)^4}{4!}(x - 1)^4$

65. $\frac{a(\ln a)^n}{n!}(x - 1)^n$

66. $f^{(n+1)}(c)\frac{(x - a)^{n+1}}{(n + 1)!}$, a and x, $x = a$, $\lim_{n \to \infty} R_n(x) = 0$

67. 0

OBJECTIVE C : Use the Remainder Estimation Theorem to estimate the truncation error when a Taylor polynomial is used to approximate a given function. Assume that the function has derivatives of all orders.

68. We will calculate $\cos \sqrt{2}$ with an error less than 10^{-6}. By Taylor's Theorem, $\cos \sqrt{2} = $ _____ $+ R_{2k}(x)$. The Remainder Estimation Theorem, with $M = $ _____, $x = $ _____, and $r = 1$ gives $|R_{2k}| \leq 1 \cdot$ _____. By trial we find

that $\frac{(\sqrt{2})^{11}}{11!} = 0.0000011337 > 10^{-6}$ and $\frac{(\sqrt{2})^{13}}{13!} = 0.0000000145$

$< 10^{-6}$. Thus, we should take $(2k + 1)$ to be at least _____,

or k to be at least 6. With an error less than 10^{-6},

$$\cos \sqrt{2} = 1 - \frac{2}{2!} + \frac{4}{4!} - \frac{8}{6!} + \cdots + \underline{\qquad}$$
$$\uparrow \text{ last term}$$

$$\approx 0.155944.$$

69. Let us determine for what values of $x > 0$ we can replace e^x by $1 + x + \left(\frac{x^2}{2}\right) + \left(\frac{x^3}{3!}\right)$ with an error of magnitude less than 5×10^{-5}. In Example 4, page 629 of the text, if $x > 0$,

$$|R_3(x)| < \underline{\qquad\qquad\qquad}.$$

We desire $|R_3(x)| < 5 \times 10^{-5}$. This is the case if $e^x|x|^4 < $ _____ or, since $x > 0$, $x + 4 \ln x < -5.116$. By calculator experimentation this inequality holds if $0 < x < 0.26$. Thus, for instance,

$$e^{0.1} = 1 + (0.1) + \frac{(0.1)^2}{2} + \frac{(0.1)^3}{6} \approx 1.10517$$

is correct to five decimal places.

OBJECTIVE D : Using the Maclaurin series for the functions e^x, $\sin x$, and $\cos x$, write the Maclaurin series for functions which are combinations of sines, cosines, exponentials, or powers of x.

70. The Maclaurin series for e^x, $\sin x$, and $\cos x$ are

$e^x = $ _____, $\sin x = $ _____, and $\cos x = $ _____.

71. Let us find the Maclaurin series for $\sin^3 x$. A trigonometric identity gives

$$\sin^3 x = \frac{1}{4}(3 \sin x - \sin 3x).$$

---

68. $1 - \frac{2}{2!} + \frac{4}{4!} - \frac{8}{6!} + \cdots + (-1)^k \frac{2^k}{(2k)!}$, 1, $\sqrt{2}$, $\frac{(\sqrt{2})^{2k+1}}{(2k+1)!}$, 13, 6, $\frac{64}{12!}$

69. $e^x \cdot \frac{x^4}{4!}$, 6×10^{-3} 70. $\sum\limits_{n=0}^{\infty} \frac{x^n}{n!}$, $\sum\limits_{n=0}^{\infty} \frac{(-1)^n x^{2n+1}}{(2n+1)!}$, $\sum\limits_{n=0}^{\infty} \frac{(-1)^n x^{2n}}{(2n)!}$

We use Maclaurin series for the terms on the right side:

$$3 \sin x = 3x - \frac{3x^3}{3!} + \frac{3x^5}{5!} - \frac{3x^7}{7!} + \cdots,$$

$$\sin 3x = \underline{\hspace{10cm}},$$

$$3 \sin x - \sin 3x = 4x^3 - 2x^5 + \frac{52}{5!} x^7 - \cdots$$

Therefore, $\sin^3 x = \underline{\hspace{6cm}}$

$$= \sum_{n=0}^{\infty} \frac{(-1)^n (3 - 3^{2n+1})}{4(2n + 1)!} x^{2n+1}.$$

72.   Euler's identity asserts that

$$e^{i\theta} = \underline{\hspace{6cm}}.$$

9.8   FURTHER CALCULATIONS WITH TAYLOR SERIES

OBJECTIVE A :   Use a suitable series to calculate a given quantity to three decimal places. Show that the remainder term does not exceed 5×10^{-4}. (Assume the quantity is the value of a function whose series expansion has been studied in this chapter of the text.)

73.   Replacing x by $-x$ in the Taylor series expansion for $\ln(1 + x)$ gives the expansion

$$\ln(1 - x) = \underline{\hspace{8cm}}$$

which is also valid for $|x| < 1$. Subtracting this result from the expansion for $\ln(1 + x)$ gives

$$\ln(1 + x) - \ln(1 - x) = \ln \frac{1 + x}{1 - x} = \underline{\hspace{6cm}}.$$

74.   Let N be a positive integer. Then

$$\ln(N + 1) = \ln N + \ln \frac{N + 1}{N}.$$

Now, solve the equation

$$\frac{1 + x}{1 - x} = \frac{N + 1}{N}$$

for x to obtain $x = \underline{\hspace{2cm}}$. Substitution into the result from Problem 73 yields

$$\ln \frac{N + 1}{N} = \underline{\hspace{6cm}}.$$

---

71. $3x - \frac{(3x)^3}{3!} + \frac{(3x)^5}{5!} - \frac{(3x)^7}{7!} + \cdots$, $x^3 - \frac{1}{2} x^5 + \frac{13}{5!} x^7 - \cdots$ 72. $\cos \theta + i \sin \theta$

73. $-x - \frac{x^2}{2} - \frac{x^3}{3} - \cdots - \frac{x^n}{n} - \cdots$, $2\left(x + \frac{x^3}{3} + \frac{x^5}{5} + \cdots + \frac{x^{2k-1}}{2k - 1} + \cdots\right)$

74. $\frac{1}{2N + 1}$, $2\left(\frac{1}{2N + 1} + \frac{1}{3(2N + 1)^3} + \frac{1}{5(2N + 1)^5} + \cdots\right)$

75.   Let's use the result of Problem 73 to calculate $\ln 2$ by setting $N = 1$. Thus,

$$\ln 2 = 2\left(\frac{1}{3} + \frac{1}{3(3)^3} + \frac{1}{3(3)^5} + \frac{1}{7(3)^7} + \frac{1}{9(3)^9} + \frac{1}{11(3)^{11}} + \cdots\right)$$

$$\approx 2(0.3333333 + 0.0123457 + 0.0008230 + 0.0000653$$
$$+ 0.0000056 + 0.0000005)$$

$$\approx \underline{\hspace{4cm}}.$$

The error satisfies

$$|R_{11}(x)| \leq \frac{1}{12} \cdot \frac{|x|^{12}}{1 - |x|} \quad \text{where} \quad x = \frac{1}{2N + 1} = \underline{\hspace{2cm}}.$$

Thus,

$$|R_{11}(x)| \leq \frac{1}{12} \cdot \frac{3}{2}\left|\frac{1}{3}\right|^{12} = 0.00000023252 < 5 \times 10^{-6}.$$

It follows that $\ln 2 \approx 0.69315$, accurate to five decimal places.

76.   Now let's calculate $\ln 0.75$ using the result in Problem 73. First use the identity

$$\ln 0.75 = \ln \frac{3}{4} = \ln \frac{3}{2} - \underline{\hspace{4cm}}.$$

We can use the calculation for $\ln 2$ obtained in Problem 75: $\ln 2 \approx 0.69315$. To obtain the first term on the right side of the previous equation we use the series

$$\ln \frac{N + 1}{N} = 2\left(\underline{\hspace{5cm}}\right) \quad \text{with} \quad N = 2.$$

Then,

$$\ln \frac{3}{2} = 2\left(\frac{1}{5} + \frac{1}{3(5)^3} + \frac{1}{5(5)^5} + \frac{1}{7(5)^7} + \cdots\right)$$

$$= 2(0.2 + 0.00266667 + 0.000064 + 0.00000183 + \cdots)$$

$$\approx 0.40547.$$

The error satisfies

$$|R_7(x)| \leq \frac{1}{8} \cdot \frac{|x|^8}{1 - |x|}, \quad \text{where} \quad x = \frac{1}{2N + 1} = \underline{\hspace{2cm}}.$$

Thus,

$$|R_7(x)| \leq \frac{5}{32} \left|\frac{1}{5}\right|^8 < 5 \times 10^{-6}.$$

It follows that

$$\ln \frac{3}{4} \approx 0.40547 - 0.69315 = -0.28768,$$

accurate to five decimal places.

---

75. 0.6931468, $\frac{1}{3}$ 76. $\ln 2$, $\frac{1}{2N + 1} + \frac{1}{3(2N + 1)^3} + \frac{1}{5(2N + 1)^5} + \cdots$, $\frac{1}{5}$

OBJECTIVE B : If f is a function having a known power series
$f(x) = \sum a_n x^n$, use the series and a calculator to
estimate the integral $\int_0^b f(x)\,dx$, assuming that b
lies within the interval of convergence.

77. Let us find $\int_0^{0.2} \cos \sqrt{x}\,dx$ accurate to five decimal places.
Now,
$$\cos x = 1 - \frac{x^2}{2!} + \frac{x^4}{4!} - \frac{x^6}{6!} + \frac{x^8}{8!} - \cdots ,$$

so the power series for $\cos \sqrt{x}$ is given by

$$\cos \sqrt{x} = \underline{\hspace{4in}} , \quad x \geq 0.$$
Thus, using term-by-term integration,

$$\int_0^{0.2} \cos \sqrt{x}\,dx = \underline{\hspace{3in}} \Big]_0^{0.2}$$

$$= 0.2 - \frac{0.04}{4} + \frac{0.008}{72} - \frac{0.0016}{2880} + \frac{0.00032}{201600} - \cdots$$

$$\approx 0.2 - 0.01 + 0.00011 - 0.00000056 + \cdots$$

Hence, $\int_0^{0.2} \cos \sqrt{x}\,dx \approx \underline{\hspace{1.5in}}$

with an error less than 5×10^{-6}.

---

77. $1 - \frac{x}{2!} + \frac{x^2}{4!} - \frac{x^3}{6!} + \frac{x^4}{8!} - \cdots$, $x - \frac{x^2}{2\cdot 2!} + \frac{x^3}{3\cdot 4!} - \frac{x^4}{4\cdot 6!} + \frac{x^5}{5\cdot 8!} - \cdots$, 0.19011

CHAPTER 9   SELF-TEST

1. Determine if each sequence $\{a_n\}$ converges or diverges. Find the limit of the sequence if it does converge.

    (a) $a_n = \sqrt{n + 1} - \sqrt{n}$

    (b) $a_n = \dfrac{1 + (-1)^n}{n\sqrt{n}}$

    (c) $a_n = \left(\dfrac{n - 0.05}{n}\right)^n$

    (d) $a_n = \dfrac{2^n}{5^{3 + 1/n}}$

2. Find the sum of each series.

    (a) $\displaystyle\sum_{n=0}^{\infty} (-1)^n \frac{3}{5^n}$

    (b) $\displaystyle\sum_{n=4}^{\infty} \frac{2}{(4n - 3)(4n + 1)}$

    (c) $\displaystyle\sum_{n=0}^{\infty} \left(\frac{5}{3^n} - \frac{2}{7^n}\right)$

    (d) $\dfrac{127}{1000} + \dfrac{127}{1000^2} + \dfrac{127}{1000^3} + \cdots + \dfrac{127}{1000^n} + \cdots$

In Problems 3-8, determine whether the given series converges or diverges. In each case, give a reason for your answer.

3. $\displaystyle\sum_{n=1}^{\infty} \frac{\sqrt{n}}{n^2 + 3}$

4. $\displaystyle\sum_{n=1}^{\infty} \frac{n!\,3^n}{10^n}$

5. $\displaystyle\sum_{n=1}^{\infty} \sin\left(\frac{n\pi - 2}{3n}\right)$

6. $\displaystyle\sum_{n=1}^{\infty} \left(\frac{n}{2n + 5}\right)^n$

7. $\displaystyle\sum_{n=1}^{\infty} \frac{1}{n + \sqrt{n}}$

8. $\displaystyle\sum_{n=1}^{\infty} \frac{\tan^{-1} n}{n^2 + 1}$

In Problems 9-12, determine whether the series are absolutely convergent, conditionally convergent, or divergent.

9. $\displaystyle\sum_{n=1}^{\infty} (-1)^{n+1} \frac{\sin n}{n^2 + 1}$

10. $\displaystyle\sum_{n=1}^{\infty} (-1)^{n+1} \frac{1}{(n + 1)^{1/n}}$

11. $\displaystyle\sum_{n=2}^{\infty} (-1)^n \frac{1}{(\ln n)^2}$

12. $\displaystyle\sum_{n=1}^{\infty} (-1)^{n+1} \frac{n + 1}{7n - 2}$

13. Estimate the magnitude of the error if the first five terms are used to approximate the series,

$$\sum_{n=1}^{\infty} (-1)^{n+1} \frac{2^n}{3^n} .$$

Sum the first five terms, and state whether your approximation underestimates or overestimates the sum of the series.

14. Find the interval of convergence for the power series

$$\sum_{n=2}^{\infty} \frac{\ln n}{n} x^n$$

15. Find the Taylor series of $f(x) = \sqrt{x}$ at $a = 9$. Do not be concerned with whether the series converges to the given function f.

16. Find the Maclaurin series for the function $f(x) = x \ln(1 + x^2)$ using series that have already been obtained in the Finney/Thomas text.

17. Use series to estimate the number $e^{-1/3}$ with an error of magnitude less than 0.001.

18. Find the first three nonzero terms in the Maclaurin series for the function $f(x) = \sec^2 x$ using the Maclaurin series for $\tan x$.

19. (Calculator) Use series and a calculator to estimate the integral

$$\int_0^{0.5} \cos x^2 \, dx$$

with an error of magnitude less than 0.0001.

SOLUTIONS TO CHAPTER 9 SELF-TEST

1. (a) $a_n = \sqrt{n + 1} - \sqrt{n} = \dfrac{(\sqrt{n + 1} - \sqrt{n})(\sqrt{n + 1} + \sqrt{n})}{(\sqrt{n + 1} + \sqrt{n})} = \dfrac{(n + 1) - n}{\sqrt{n + 1} + \sqrt{n}}$

$$= \frac{1}{\sqrt{n + 1} + \sqrt{n}} \to 0 \quad \text{as} \quad n \to \infty$$

(b) $\sqrt[n]{n} \to 1$, but $1 + (-1)^n$ alternates back and forth between 0 and 2. Thus, for n large, a_n alternates between numbers very close to 2 and 0; hence the sequence diverges.

(c) $a_n = \left(\dfrac{n - 0.05}{n}\right)^n = \left(1 + \dfrac{-0.05}{n}\right)^n \to e^{-0.05} \approx 0.951$.

(d) $5^{3 + 1/n} = 125 \sqrt[n]{5} \to 125$, but $2^n \to +\infty$. Therefore, the sequence $\{a_n\}$ is unbounded and diverges.

2. (a) $\displaystyle\sum_{n=0}^{\infty} (-1)^n \frac{3}{5^n} = \sum_{n=0}^{\infty} 3\left(-\frac{1}{5}\right)^n = \frac{3}{1 + \frac{1}{5}} = \frac{5}{2}$.

(b) Using the partial fraction decomposition,

$$\frac{2}{(4k - 3)(4k + 1)} = \frac{1}{2}\left(\frac{1}{4k - 3}\right) - \frac{1}{2}\left(\frac{1}{4k + 1}\right), \quad \text{we write the}$$

partial sum

$$s_k = \sum_{n=4}^{k} \frac{2}{(4n - 3)(4n + 1)}$$

as

$$s_k = \frac{1}{2}\left(\frac{1}{13} - \frac{1}{17}\right) + \frac{1}{2}\left(\frac{1}{17} - \frac{1}{21}\right) + \frac{1}{2}\left(\frac{1}{21} - \frac{1}{25}\right) + \cdots$$

$$+ \frac{1}{2}\left(\frac{1}{4k - 3} - \frac{1}{4k + 1}\right).$$

Thus,

$$s_k = \frac{1}{2}\left(\frac{1}{13} - \frac{1}{4k + 1}\right) \to \frac{1}{26} \quad \text{as} \quad k \to \infty$$

so that

$$\sum_{n=4}^{\infty} \frac{2}{(4n - 3)(4n + 1)} = \frac{1}{26}.$$

(c) $\displaystyle\sum_{n=0}^{\infty}\left(\frac{5}{3^n} - \frac{2}{7^n}\right) = \sum_{n=0}^{\infty}\frac{5}{3^n} - \sum_{n=0}^{\infty}\frac{2}{7^n} = \frac{5}{1 - \frac{1}{3}} - \frac{2}{1 - \frac{1}{7}} = \frac{31}{6}$.

(d) $\displaystyle\sum_{n=1}^{\infty} 127\left(\frac{1}{1000}\right)^n = \sum_{n=0}^{\infty} 127\left(\frac{1}{1000}\right)^n - 127 = \frac{127}{1 - \frac{1}{1000}} - 127$

$$= \frac{127{,}000 - 126{,}873}{999} = \frac{127}{999}.$$

3. $\dfrac{\sqrt{n}}{n^2 + 3} < \dfrac{\sqrt{n}}{n^2} = \dfrac{1}{n^{3/2}}$ so that $\displaystyle\sum_{n=1}^{\infty} \dfrac{\sqrt{n}}{n^2 + 3}$ <u>converges</u> by comparison with the convergent p-series for $p = \dfrac{3}{2}$.

4. Using the ratio test, $\displaystyle\lim_{n \to \infty} \dfrac{(n + 1)! \, 3^{n+1}}{10^{n+1}} \cdot \dfrac{10^n}{n! \, 3^n} = \lim_{n \to \infty} \dfrac{(n + 1)3}{10}$ $= \infty$.   Thus, $\displaystyle\sum_{n=1}^{\infty} \dfrac{n! \, 3^n}{10^n}$ <u>diverges</u> by the ratio test.

5. $\displaystyle\lim_{n \to \infty} \sin\left(\dfrac{n\pi - 2}{3n}\right) = \lim_{n \to \infty} \sin\left(\dfrac{\pi}{3} - \dfrac{2}{3n}\right) = \sin\dfrac{\pi}{3} = \dfrac{\sqrt{3}}{2} \neq 0$,   so the series $\displaystyle\sum_{n=1}^{\infty} \sin\left(\dfrac{n\pi - 2}{3n}\right)$ <u>diverges</u> by the nth-term test for divergence.

6. If $a_n = \left(\dfrac{n}{2n + 5}\right)^n$,   then $\sqrt[n]{a_n} = \dfrac{n}{2n + 5} \to \dfrac{1}{2}$.   Thus, the series $\displaystyle\sum_{n=1}^{\infty} \left(\dfrac{n}{2n + 5}\right)^n$ <u>converges</u> by the root test.

7. $\dfrac{1}{n + \sqrt{n}} > \dfrac{1}{n + n} = \dfrac{1}{2n}$ so that $\displaystyle\sum_{n=1}^{\infty} \dfrac{1}{n + \sqrt{n}}$ <u>diverges</u> by comparison to the divergent series $\displaystyle\sum_{n=1}^{\infty} \dfrac{1}{2n}$.

8. $\displaystyle\int_{1}^{\infty} \dfrac{\tan^{-1}x \, dx}{x^2 + 1} = \lim_{b \to \infty} \dfrac{1}{2}\left(\tan^{-1}x\right)^2 \Big]_{1}^{b} = \lim_{b \to \infty} \dfrac{1}{2}\left(\tan^{-1}b\right)^2 - \dfrac{1}{2}\tan^{-1}1$
$= \dfrac{1}{2}\left(\dfrac{\pi}{2}\right)^2 - \dfrac{1}{2}\left(\dfrac{\pi}{4}\right)$

Therefore, the improper integral converges, so the original series $\displaystyle\sum_{n=1}^{\infty} \dfrac{\tan^{-1}n}{n^2 + 1}$ <u>converges</u> by the integral test.

9. $\left|(-1)^{n+1} \dfrac{\sin n}{n^2 + 1}\right| \leq \dfrac{1}{n^2 + 1}$,   so the original series

$\displaystyle\sum_{n=1}^{\infty} (-1)^{n+1} \dfrac{\sin n}{n^2 + 1}$ <u>converges</u> <u>absolutely</u> by the comparison test.

10. Since $\dfrac{1}{n^2} < \dfrac{1}{n + 1} < \dfrac{1}{n}$,   it follows that

$\left(-\dfrac{1}{n\sqrt{n}}\right)\left(-\dfrac{1}{n\sqrt{n}}\right) < \dfrac{1}{n\sqrt{n + 1}} < \dfrac{1}{n\sqrt{n}}$.   Thus, $\displaystyle\lim_{n \to \infty} \dfrac{1}{n\sqrt{n + 1}} = 1$ so

the original series $\displaystyle\sum_{n=1}^{\infty} (-1)^{n+1} \dfrac{1}{(n + 1)^{1/n}}$ <u>diverges</u> by the nth-term test.

11. $\displaystyle\lim_{n \to \infty} \dfrac{1}{(\ln n)^2} = 0$,   and $\dfrac{1}{[\ln (n + 1)]^2} < \dfrac{1}{(\ln n)^2}$ because $y = \ln x$ is an increasing function of x.   Therefore the alternating

series $\sum_{n=2}^{\infty} (-1)^n \frac{1}{(\ln n)^2}$ converges by the Alternating Series

Theorem. However, since $\ln n < \sqrt{n}$ implies $\frac{1}{(\ln n)^2} > \frac{1}{n}$ if

$n \geq 2$, series of absolute values $\sum_{n=2}^{\infty} \frac{1}{(\ln n)^2}$ diverges by

comparison with the divergent harmonic series. Therefore,

$\sum_{n=2}^{\infty} (-1)^n \frac{1}{(\ln n)^2}$ is <u>conditionally convergent</u>.

12. $\lim_{n \to \infty} \frac{n + 1}{7n - 2} = \frac{1}{7}$ so that $\sum_{n=1}^{\infty} (-1)^{n+1} \frac{n + 1}{7n - 2}$ <u>diverges</u> by the nth-
term test.

13. $\sum_{n=1}^{\infty} (-1)^{n+1} \frac{2^n}{3^n} \approx \frac{2}{3} - \frac{4}{9} + \frac{8}{27} - \frac{16}{81} + \frac{32}{243} \approx 0.4527$ with an error of

magnitude less than $2^6/3^6 < 0.0878$. Since the sign of the first
unused term is negative, the sum 0.4527 overestimates the value
of the series. In fact, the given geometric series sums to 0.4.

14. Using the ratio test,

$$\lim_{n \to \infty} \frac{\ln (n + 1) |x|^{n+1}}{(n + 1)} \cdot \frac{n}{\ln n |x|^n}$$

$$= \lim_{n \to \infty} \frac{\ln (n + 1)}{\ln n} \cdot \frac{n + 1}{n} |x| = |x|.$$

Thus, the given power series converges absolutely for
$|x| < 1$ and diverges for $|x| > 1$. Test the endpoints of the
interval:

For $x = 1$, the power series is $\sum_{n=2}^{\infty} \frac{\ln n}{n}$. Now

$\int_2^{\infty} \frac{\ln x}{x} dx = \lim_{b \to \infty} \frac{1}{2} (\ln x)^2 \big]_2^b = +\infty$ diverges, so the series

$\sum_{n=2}^{\infty} \frac{\ln n}{n}$ is divergent by the integral test.

For $x = -1$, the power series is $\sum_{n=2}^{\infty} \frac{(-1)^n \ln n}{n}$. Since

$0 \leq \lim_{n \to \infty} \frac{\ln n}{n} \leq \lim_{n \to \infty} \frac{\sqrt{n}}{n} = 0$, and $\frac{d}{dx} \left(\frac{\ln x}{x} \right) = \frac{1 - \ln x}{x^2} < 0$ for

$x \geq 3$ implies that $\frac{\ln (n + 1)}{n + 1} < \frac{\ln n}{n}$, the alternating series

$\sum_{n=2}^{\infty} \frac{(-1)^n \ln n}{n}$ converges by the Alternating Series Theorem.

Therefore, the power series $\sum_{n=2}^{\infty} \frac{\ln n}{n} x^n$ converges for all x

satisfying $-1 \leq x < 1$.

15. We calculate the derivatives of $f(x) = \sqrt{x}$, and evaluate f and these derivatives at $a = 9$:

$$f(x) \quad = \sqrt{x} \qquad\qquad\qquad f(9) \quad = 3$$

$$f'(x) \quad = \tfrac{1}{2}x^{-1/2} \qquad\qquad f'(9) \quad = \tfrac{1}{6}$$

$$f^{(2)}(x) = (-1)\left(\tfrac{1}{2}\right)\left(\tfrac{1}{2}\right)x^{-3/2} \qquad f^{(2)}(9) = -\tfrac{1}{108}$$

$$f^{(3)}(x) = (-1)^2\left(\tfrac{1}{2}\right)\left(\tfrac{1}{2}\right)\left(\tfrac{3}{2}\right)x^{-5/2} \qquad f^{(3)}(9) = \tfrac{1}{648}$$

$$f^{(4)}(x) = (-1)^3\,\frac{3\cdot 5}{2^4}\,x^{-7/2} \qquad f^{(4)}(9) = -\tfrac{5}{11664}$$

$$\vdots \qquad\qquad\qquad\qquad \vdots$$

$$f^{(k)}(x) = (-1)^{k+1}\,\frac{3\cdot 5\cdots(2k-3)}{2^k}\,x^{-(2k-1)/2}$$

$$f^{(k)}(9) = (-1)^{k+1}\,\frac{3\cdot 5\cdots(2k-3)}{2^k\,3^{2k-1}}$$

Therefore, the Taylor series for $f(x) = \sqrt{x}$ at $a = 9$ is

$$3 + \tfrac{1}{6}(x-9) - \tfrac{1}{216}(x-9)^2 + \ldots + (-1)^{k+1}\,\frac{3\cdot 5\cdots(2k-3)}{2^k\,3^{2k-1}\cdot k!}\,(x-9)^k + \ldots$$

16.
$$\ln(1+x) = x - \frac{x^2}{2} + \frac{x^3}{3} - \frac{x^4}{4} + \ldots\,, \qquad -1 < x \le 1$$

$$\ln(1+x^2) = x^2 - \frac{x^4}{2} + \frac{x^6}{3} - \frac{x^8}{4} + \ldots\,, \qquad -1 < x \le 1$$

$$x\ln(1+x^2) = x^3 - \frac{x^5}{2} + \frac{x^7}{3} - \frac{x^9}{4} + \ldots\,, \qquad -1 < x \le 1$$

or, in closed form, $\;x\ln(1+x^2) = \sum\limits_{n=0}^{\infty} (-1)^n\,\frac{1}{n+1}\,x^{2n+3}$, valid for all x satisfying $-1 < x \le 1$.

17.
$$e^{-1/3} = 1 - \frac{1}{3} + \frac{(-1/3)^2}{2!} + \frac{(-1/3)^3}{3!} + \frac{(-1/3)^4}{4!} + \ldots$$

By trial, $\dfrac{(-1/3)^4}{4!} < 0.00052$ and $\dfrac{(1/3)^3}{3!} > 0.001$.

Since the series is an alternating series,

$$e^{-1/3} \approx 1 - \frac{1}{3} + \frac{1/9}{2!} - \frac{1/27}{3!} = 0.71605 \quad \text{with an error in magnitude}$$

less than 0.00052.

18. The Maclaurin series for $\tan x$, through the first three nonzero terms, is

$$\tan x = x + \frac{x^3}{3} + \frac{2x^5}{15} + \ldots$$

Hence, $\sec^2 x = \dfrac{d}{dx}\tan x = 1 + x^2 + \tfrac{2}{3}x^4 + \ldots\,.$

19.   The Maclaurin series for $\cos x^2$ is

$$\cos x^2 = 1 - \frac{x^4}{2!} + \frac{x^8}{4!} - \frac{x^{12}}{6!} + \cdots + (-1)^k \frac{x^{4k}}{(2k)!} + \cdots$$

Hence,

$$\int_0^{0.5} \cos x^2 \, dx = x - \frac{x^5}{5 \cdot 2!} + \frac{x^9}{9 \cdot 4!} - \frac{x^{13}}{13 \cdot 6!} + \cdots \Big]_0^{0.5}$$

$$\approx 0.5 - 0.00313 + 0.0000090 - \cdots = 0.49687,$$

with an error in magnitude less than 0.000009 because the
series is an alternating series.

NOTES.

CHAPTER 10 PLANE CURVES AND POLAR COORDINATES

10.1 CONIC SECTIONS AND QUADRATIC EQUATIONS

$\boxed{\text{OBJECTIVE A}}$: Given an equation of an ellipse $Ax^2 + Cy^2 = F$, where A, C and F are positive numbers, put the equation in standard form and find the ellipse's eccentricity. Sketch the ellipse and include the foci in your sketch.

1. The standard equation for an ellipse centered at the origin with foci on the x-axis is

    _____ , $a > b$.

    The foci are located at the points _____ where $c =$ _____. The points $(\pm a, 0)$ are called the _____ of the ellipse, and a is the _____ axis.

2. If the foci lie on the y-axis, the standard equation for an ellipse centered at the origin is

    _____ , $a > b$.

    The points $(0, \pm a)$ are the _____ and the foci are _____ where $c = \sqrt{a^2 - b^2}$.

3. The ratio

    $$\frac{\text{Distance between foci}}{\text{Distance between vertices}} = \frac{c}{a}$$

    is called the _____ of the ellipse. If $c = 0$ the ellipse takes the shape of a _____. As c increases, the ellipse tends to _____, degenerating into a _____ when $c = a$.

4. In an elliptical mirror, light is reflected from _____ .

---

1. $\dfrac{x^2}{a^2} + \dfrac{y^2}{b^2} = 1$, $(\pm c, 0)$, $\sqrt{a^2 - b^2}$, vertices, semimajor

2. $\dfrac{x^2}{b^2} + \dfrac{y^2}{a^2} = 1$, vertices, $(0, \pm c)$

3. eccentricity, circle, flatten out, line segment

4. one focus to the other

5. The equation $16x^2 + 5y^2 = 80$ represents an ellipse. In standard form it is written

_____ .

In this case c = _____ ≈ 3.32, and the foci are located at the points _____ . The vertices are
_____ . The eccentricity is

e = _____ .

Sketch the ellipse in the coordinate system at the right, and include the foci in your sketch.

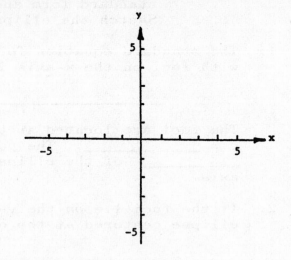

OBJECTIVE B : Given an equation of a hyperbola $Ax^2 - Cy^2 = F$ or $Cy^2 - Ax^2 = F$, where A, C and F are positive numbers, put the equation in standard form and find the hyperbola's eccentricity and asymptotes. Sketch the hyperbola, including the asymptotes and foci in your sketch.

6. The standard equation for a hyperbola centered at the origin with foci on the x-axis is

_____ .

---

5. $\dfrac{x^2}{5} + \dfrac{y^2}{16} = 1$, $\sqrt{16 - 5}$,

$(0, \pm\sqrt{11})$, $(0, \pm 4)$, $\dfrac{\sqrt{11}}{4}$

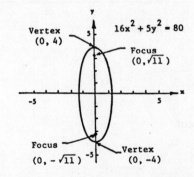

The foci are located at the points _____ where
c = _____. The points $(\pm a, 0)$ are called the
_____ of the hyperbola. The asymptotes are straight lines
passing through the origin with equations _____.
Note that in the case of the hyperbola it is not necessary for
a > b or b > a. It is also permissible for a = b.

7. If the foci lie on the y-axis, the standard equation for a
    hyperbola centered at the origin is

    _____ .

    The foci are located at the points $(0, \pm c)$ where
    c = _____. The points _____ are the vertices, and
    the asymptotes are given by the equations _____.

8. The eccentricity of a hyperbola is always given by

    e = _____ .

    For a hyperbola, it is always true that e _____ 1.

9. Light directed toward one focus of a hyperbolic mirror is
    reflected _____.

10. The equation $16y^2 - 5x^2 = 80$ represents a hyperbola. In
    standard form it is written

    _____ .

    In this case c = _____ ≈ 4.58, and the foci are located
    at the points _____. The vertices are _____,
    and the asymptotes are given by _____. The
    eccentricity is e = _____.

---

6. $\dfrac{x^2}{a^2} - \dfrac{y^2}{b^2} = 1$, $(\pm c, 0)$, $\sqrt{a^2 + b^2}$, vertices, $y = \pm \dfrac{b}{a} x$

7. $\dfrac{y^2}{a^2} - \dfrac{x^2}{b^2} = 1$, $\sqrt{a^2 + b^2}$, $(0, \pm a)$, $y = \pm \dfrac{a}{b} x$

8. $\dfrac{c}{a}$ or $\dfrac{\sqrt{a^2 + b^2}}{a}$, e > 1

9. toward the other focus

Sketch the hyperbola in the
coordinate system at the
right, including the foci
and asymptotes in your
sketch.

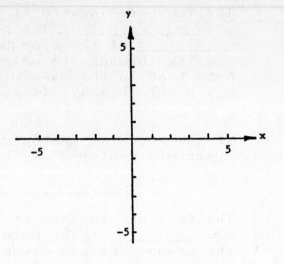

10.2   THE GRAPHS OF QUADRATIC EQUATIONS IN x AND y

11.   Any second degree equation in x and y represents a circle,
      parabola, ellipse, or hyperbola (although it may degenerate).
      To find the curve given its equation
      $$Ax^2 + Bxy + Cy^2 + Dx + Ey + F = 0,$$
      (1)   First rotate axes, to force B = 0, through an angle α
            satisfying $\cot 2\alpha = $ _____ .
      (2)   Next, _____ axes by completing the squares (if
            necessary) to reduce the equation to a standard form.

OBJECTIVE A :   Given an equation of the form $Ax^2 + Bxy + Cy^2 + F = 0$,
                transform the equation by a rotation of axes into an
                equation that has no cross-product term.   Then identify
                the graph of the equation.

---

10. $\frac{y^2}{5} - \frac{x^2}{16} = 1$, $\sqrt{5 + 16}$, $(0, \pm\sqrt{21})$,

      $(0, \pm\sqrt{5})$, $y = \pm\frac{\sqrt{5}}{4}x$,

      $\frac{\sqrt{21}}{\sqrt{5}} \approx 2.05$

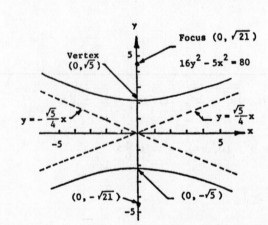

11. $\frac{A - C}{B}$, translate

12.   Consider the equation $5x^2 - 2\sqrt{3}xy + 7y^2 = 6$. The required angle of rotation to eliminate the xy term satisfies $\cot 2\alpha = \dfrac{5 - 7}{\underline{\hspace{1cm}}} = \underline{\hspace{1.5cm}}$; hence $2\alpha = \underline{\hspace{1cm}}$ radians, or $\alpha = \underline{\hspace{1.5cm}}$. Thus, $\sin \alpha = \underline{\hspace{1cm}}$ and $\cos \alpha = \underline{\hspace{1.5cm}}$. The equations for the rotation of axes are $x = \dfrac{\sqrt{3}}{2}x' - \dfrac{1}{2}y'$ and $\underline{\hspace{4cm}}$. Substitution for x and y into the given second-degree equation gives

$$5\left(\tfrac{3}{4}x'^2 - \underline{\hspace{1.5cm}} + \tfrac{1}{4}y'^2\right) - 2\sqrt{3}\left(\tfrac{\sqrt{3}}{4}x'^2 + \underline{\hspace{1.5cm}} - \tfrac{\sqrt{3}}{4}y'^2\right)$$
$$+ 7\left(\tfrac{1}{4}x'^2 + \underline{\hspace{1.5cm}} + \tfrac{3}{4}y'^2\right) = 6 .$$

Simplifying algebraically,

$$\left(\tfrac{15}{4} - \underline{\hspace{1.5cm}} + \tfrac{7}{4}\right)x'^2 + \left(\underline{\hspace{0.8cm}} + \tfrac{3}{2} + \tfrac{21}{4}\right)y'^2 = 6, \text{ or}$$

$\underline{\hspace{4cm}}$. This is an equation of $\underline{\hspace{3cm}}$.

OBJECTIVE B: Given a second degree equation of the form $Ax^2 + Bxy + Cy^2 + Dx + Ey + F = 0$, use the discriminant to classify it as representing a circle, an ellipse, a parabola, or a hyperbola.

13.   The <u>discriminant</u> is the invariant expression $\underline{\hspace{3cm}}$. The discriminant is not changed by a rotation of axes.

14.   If the discriminant is positive, the equation represents $\underline{\hspace{2.5cm}}$.

15.   If the discriminant is zero, the equation represents $\underline{\hspace{2.5cm}}$.

16.   If the discriminant is negative, the equation represents $\underline{\hspace{2.5cm}}$.

17.   In order that the equation represent a circle, it is necessary that the discriminant be $\underline{\hspace{2cm}}$ and that $\underline{\hspace{2.5cm}}$.

18.   Consider the equation given by
$$2x^2 - 4xy - y^2 + 20x - 2y + 17 = 0.$$
The discriminant is $B^2 - 4AC = 16 - (\underline{\hspace{1.5cm}}) = \underline{\hspace{1.5cm}}$. Thus, the equation represents $\underline{\hspace{3cm}}$.

---

12. $-2\sqrt{3}$, $\dfrac{1}{\sqrt{3}}$, $\dfrac{\pi}{3}$, $\dfrac{\pi}{6}$, $\dfrac{1}{2}$, $\dfrac{\sqrt{3}}{2}$, $y = \tfrac{1}{2}x' + \tfrac{\sqrt{3}}{2}y'$, $\dfrac{\sqrt{3}}{2}x'y'$, $\tfrac{1}{2}x'y'$, $\dfrac{\sqrt{3}}{2}x'y'$, $\dfrac{3}{2}$,

   $\dfrac{5}{4}$, $4x'^2 + 8y'^2 = 6$, an ellipse

13. $B^2 - 4AC$, $A + C$ 14. a hyperbola   15. a parabola   16. an ellipse

17. negative, $A = C$ 18. -8, 24, a hyperbola

19. For $4x^2 - 12xy + 9y^2 - 52x + 26y + 81 = 0$ the discriminant is $B^2 - 4AC =$ _____ $- 4(36) =$ _____. Thus, the equation represents _____.

10.3   PARAMETRIC EQUATIONS FOR PLANE CURVES

OBJECTIVE A :   Given parametric equations $x = f(t)$ and $y = g(t)$ for the motion of a particle in the xy-plane, eliminate the parameter t to find a Cartesian equation for the particle's path. Graph the Cartesian equation.

20. Consider the curve given by the parametric equations $x = t - 2$, $y = 2t + 3$, $-\infty < t < \infty$. Complete the following table providing some of the points $P(x,y)$ on the curve:

t	-2	-1	0	1	2	3
x	-4					
y	-1					

To eliminate the parameter t, note that $t = x + 2$. Substitution for t in the parametric equation for y gives $y =$ _____. This is a Cartesian equation for a _____ with slope $m =$ _____ and y-intercept $b =$ _____.
Sketch the curve in the coordinate system to the right.

21. For the curve given by the parametric equations $x = e^t$ and $y = e^{-t}$, $-\infty < t < \infty$, complete the following table:

t	-2	-1	0	1	2	3
x						
y						

---

19. 144, 0, a parabola

20.

t	-2	-1	0	1	2	3
x	-4	-3	-2	-1	0	1
y	-1	1	3	5	7	9

$2x + 7$, line, 2, 7

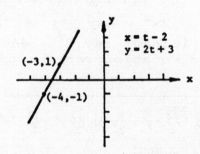

$x = t - 2$
$y = 2t + 3$
(-3,1)
(-4,-1)

To eliminate the parameter t, notice that xy = _____. This equation describes a _____. Sketch the graph in the coordinate system at the right.

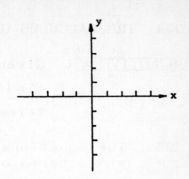

22. For the curve given parametrically in Problem 21, notice that x and y are always positive. Are the parametric equations and the cartesian equation coextensive? _____, because x and y can both be _____ in the cartesian equation xy = 1.

OBJECTIVE B : Find parametric equations for a curve described geometrically, or by an equation, in terms of some specified or arbitrary parameter.

23. Find parametric equations for the circle with center C(-2,3) and radius r = $\sqrt{2}$.

Solution. An equation of the circle is

$(x + 2)^2$ + _____ = _____, or _____ = 1.

This suggests the substitutions $\dfrac{x + 2}{\sqrt{2}}$ = sin θ and $\dfrac{y - 3}{\sqrt{2}}$ =

_____. Hence, parametric equations for the circle are

x = _____ and y = _____, 0 ≤ θ ≤ 2π.

24. Find parametric equations for the line in the plane through the point (a,b) with slope m, where the parameter t is the change x - a.

Solution. For any point P(x,y) on the line,

y - b = m(_____) = _____. Thus, x = _____ and y = _____ give parametric equations of the line in terms of the specified parameter t.

21.

t	-2	-1	0	1	2	3
x	.14	.37	1	2.7	7.4	20
y	7.4	2.7	1	.37	.14	.05

(approximate values)

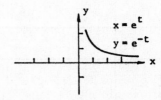

xy = 1, hyperbola

22. No, negative

23. $(y - 3)^2$, 2, $\left(\dfrac{x + 2}{\sqrt{2}}\right)^2 + \left(\dfrac{y - 3}{\sqrt{2}}\right)^2$, cos θ, $\sqrt{2}$ sin θ - 2, 3 + $\sqrt{2}$ cos θ

24. x - a, mt, a + t, b + mt

10.4   THE CALCULUS OF PARAMETRIC EQUATIONS

OBJECTIVE A : Given parametric equations $x = f(t)$ and $y = g(t)$, find $\frac{dy}{dx}$ in terms of $\frac{dy}{dt}$ and $\frac{dx}{dt}$. Find $\frac{d^2y}{dx^2}$ in terms of t.

25.   The equations $x = f(t)$ and $y = g(t)$, which express x and y in terms of t, are called _____ equations. The variable t is called a _____. From the chain rule, the derivative $\frac{dy}{dx}$ is given by $\frac{dy}{dx}$ = _____.

26.   Let $x = t^2 - 1$ and $y = \frac{1}{t}$. Then $\frac{dx}{dt}$ = _____ and $\frac{dy}{dt}$ = _____. It follows that

$$y' = \frac{dy}{dx} = \text{_____} \; .$$

To calculate the second derivative $\frac{d^2y}{dx^2}$ we first find

$$\frac{dy'}{dt} = \text{_____} \; .$$

Then,

$$\frac{d^2y}{dx^2} = \frac{dy'/dt}{\text{\_\_\_\_\_}} = \text{_____} \; .$$

27.   If $x = t^2$ and $y = t^2 - 2t$, then $\frac{dx}{dt}$ = _____ and $\frac{dy}{dt}$ = _____. Thus,

$$y' = \frac{dy}{dx} = \text{_____} \; .$$

Next, $\frac{dy'}{dt}$ = _____ so that

$$\frac{d^2y}{dx^2} = \frac{}{dx/dt} = \frac{}{2t} = \text{_____} \; .$$

When $t = 2$, $x = $ _____, $y = $ _____ and $\frac{dy}{dx} = $ _____. Thus an equation of the line tangent to the curve at $(4,0)$ is

_____ .

---

25.   parametric, parameter, $\dfrac{dy/dt}{dx/dt}$

26. $2t$, $-\dfrac{1}{t^2}$, $-\dfrac{1}{2t^3}$, $\dfrac{3}{2}t^{-4}$, $\dfrac{dx}{dt}$, $\dfrac{3}{4t^5}$

27. $2t$, $2t - 2$, $1 - \dfrac{1}{t}$, $\dfrac{dy'}{dt}$, $\dfrac{1}{t^2}$, $\dfrac{1}{2t^3}$, 4, 0, $\dfrac{1}{2}$, $y = \dfrac{1}{2}(x - 4)$ or $2y - x + 4 = 0$

[OBJECTIVE B]: Find the length of a curve specified parametrically by continuously differentiable equations $x = f(t)$ and $y = g(t)$ over a given interval $a \leq t \leq b$.

28. To compute the length of the curve given by $x = t^2 \cos t$ and $y = t^2 \sin t$ for $0 \leq t \leq 1$, we first calculate $\frac{dx}{dt}$ and $\frac{dy}{dt}$.

$\frac{dx}{dt} = $ _____ ; $\frac{dy}{dt} = $ _____ so that

$\left(\frac{dx}{dt}\right)^2 + \left(\frac{dy}{dt}\right)^2 = t^2[(2 \cos t - t \sin t)^2 + ($_____$)^2]$

$= t^2[4 \cos^2 t - 4t \cos t \sin t + t^2 \sin^2 t$

$+ ($_____$)]$

$= t^2[4 + $_____$].$

Hence the arc length is given by,

$s = \int_0^1 \sqrt{\left(\frac{dx}{dt}\right)^2 + \left(\frac{dy}{dt}\right)^2}\ dt = \int_0^1$ _____ dt

$= $ _____ $]_0^1 = \frac{1}{3}\ ($_____$) \approx 1.0601$ units.

[OBJECTIVE C]: Find the area of a surface generated by rotating the arc of a curve specified parametrically by equations $x = f(t)$ and $y = g(t)$ over $a \leq t \leq b$ about an indicated axis. (Again, f and g are assumed to have continuous derivatives.)

29. Find the surface area generated when the arc $x = 2t$ and $y = \sqrt{2}t^2$ from $t = 0$ to $t = 2$ is rotated about the y-axis. <u>Solution</u>. We write $dS = 2\pi$ ____ ds, since the rotation occurs about the y-axis. Next, we calculate the derivatives $\frac{dx}{dt} = $ ____ and $\frac{dy}{dt} = $ ____ so that the arc length differential is given by

$$ds = \sqrt{\underline{\hspace{2cm}}}\ dt = 2\sqrt{\underline{\hspace{2cm}}}\ dt.$$

Therefore, the surface area is

$S = 2\pi \int_0^2$ _____ $dt = \frac{4\pi}{3}\ ($_____$)]_0^2$

$= \frac{4\pi}{3}\ ($_____$) = $ _____ ≈ 108.90854 sq. units.

---

28. $2t \cos t - t^2 \sin t$, $2t \sin t + t^2 \cos t$, $2 \sin t + t \cos t$, $4 \sin^2 t + 4t \cos t \sin t + t^2 \cos^2 t$,

t^2, $t\sqrt{4 + t^2}$, $\frac{1}{3}\left(4 + t^2\right)^{3/2}$, $5\sqrt{5} - 8$

29. x, 2, $2\sqrt{2}t$, $\left(\frac{dx}{dt}\right)^2 + \left(\frac{dy}{dt}\right)^2$, $2t^2 + 1$, $4t\sqrt{2t^2 + 1}$, $\left(2t^2 + 1\right)^{3/2}$, $27 - 1$, $\frac{104\pi}{3}$

10.5   POLAR COORDINATES

OBJECTIVE A :   Given a point P in polar coordinates (r, θ), give
the Cartesian coordinates (x, y) of P.

30.   The polar and Cartesian coordinates are related by the
equations x = _____ and y = _____.

31.   If P is the point $(-2, \frac{\pi}{6})$ in polar coordinates, then
x = _____ and y = _____ so that P can be
expressed in Cartesian coordinates by (____,____).

32.   For $P = (-2, -\frac{\pi}{6})$ in polar coordinates, x = _____ and
y = _____ so that P = (____,____) in Cartesian
coordinates.

OBJECTIVE B :   Graph the points $P(r, \theta)$ whose polar coordinates
satisfy a given equation, inequality or inequalities.

33. $\theta = -\frac{\pi}{4}, \quad -2 \le r$
Sketch the graph at the right.

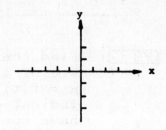

34. $r = 2, \quad -\frac{3\pi}{4} < \theta \le \frac{\pi}{6}$
Sketch the graph at the right.

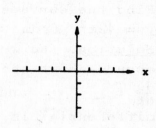

---

30. $r \cos \theta, \ r \sin \theta$ 31. $-2 \cos \frac{\pi}{6}, \ -2 \sin \frac{\pi}{6}, \ (-\sqrt{3}, -1)$ 32. $-2 \cos \left(-\frac{\pi}{6}\right), \ -2 \sin \left(-\frac{\pi}{6}\right), \ (-\sqrt{3}, 1)$

33.   34.

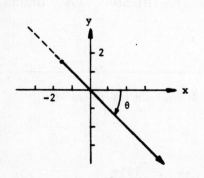

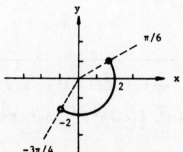

OBJECTIVE C : Given a simple equation in polar coordinates, replace it by an equivalent equation in Cartesian coordinates and sketch the graph.

35. Consider the equation $r = -3 \sec \theta$. Then, $-3 = r$ _____, or since $x =$ _____ the equation is equivalent to _____. This is an equation of a _____ line 3 units to the left of the _____ axis.

36. Consider the equation $r \sin (\theta - \frac{\pi}{3}) = \frac{1}{2}$. By the trigonometric summation identities, $\sin (\theta - \frac{\pi}{3}) = \sin \theta \cos \frac{\pi}{3} -$ _____ = _____.

    Therefore, the polar equation can be written
    $$\frac{1}{2} = \frac{1}{2} r \sin \theta - \underline{\hspace{2cm}} = \frac{1}{2} y - \underline{\hspace{1.5cm}}.$$

    Simplifying algebraically, $y =$ _____. This is an equation of a line with slope $m =$ _____ and y-intercept $b =$ _____.

37. Suppose $r = \tan \theta \sec \theta$. Then, $r = \sin \theta \cdot$ _____ or $r = \frac{\sin \theta}{\underline{\hspace{0.5cm}}}$. Equivalently, $\sin \theta =$ _____ or

    $r \sin \theta =$ _____. In terms of Cartesian coordinates the equation becomes _____, which is readily recognized as an equation of _____.

OBJECTIVE D : Given an equation in polar coordinates, replace it by an equivalent Cartesian equation and identify the graph.

38. Given the equation $r = \dfrac{1}{3 \cos \theta + 2 \sin \theta}$, clear fractions and

    obtain $3r \cos \theta +$ _____ $= 1$. Next, substitute $x = r \cos \theta$ and $y =$ _____ to obtain $3x +$ _____ $= 1$. This is an equation of a straight line with slope $m =$ _____ and y-intercept $b =$ _____.

39. The equation $r^2 - 5r + 4 = 0$ factors into $(r-4)($ _____ $) = 0$. Thus, $r = 4$ or $r =$ _____. The graph is two concentric circles, one of radius 4 and the other of radius 1, centered at the origin.

---

35. $\cos \theta$, $r \cos \theta$, $x = -3$, vertical, y

36. $\cos \theta \sin \frac{\pi}{3}$, $\frac{1}{2} \sin \theta - \frac{\sqrt{3}}{2} \cos \theta$, $\frac{\sqrt{3}}{2} r \cos \theta$, $\frac{\sqrt{3}}{2} x$, $\sqrt{3}x + 1$, $\sqrt{3}$, 1

37. $\sec^2 \theta$, $\cos^2 \theta$, $r \cos^2 \theta$, $r^2 \cos^2 \theta$, $y = x^2$, a parabola

38. $2r \sin \theta$, $r \sin \theta$, $2y$, $-\frac{3}{2}$, $\frac{1}{2}$

39. $r - 1$, 1

40.   Consider $r^2 = 2 \csc 2\theta$. Then, $r^2 \sin 2\theta =$ _____. Now, $\sin 2\theta = 2$ _____, so the equation becomes $2r \sin \theta \cdot$ _____ $= 2$ or _____ $= 2$. That is, $xy = 1$ which is an equation of _____ with center _____.

41.   Given the equation $r = \dfrac{3}{1 - 2 \cos \theta}$, clear fractions to obtain $r -$ _____ $= 3$; or substituting $x = r \cos \theta$, $r =$ _____. Hence, $r^2 =$ _____ $= 9 + 12x +$ _____. Since $r^2 = x^2 + y^2$ this last equation simplifies to $y^2 = 9 + 12x +$ _____ or $y^2 = 3(x + 2)^2 + ($_____$)$. Therefore, $(x + 2)^2 -$ _____ $= 1$. This is an equation of a hyperbola.

42.   For $r = 2 \sin \left(\theta + \frac{\pi}{4}\right)$ we can expand the right side by the summation formula for the sine: $r = 2 \sin \theta \cos \frac{\pi}{4} +$ _____ or $r =$ _____. Hence, $r^2 = \sqrt{2} r \sin \theta +$ _____. Since $x^2 + y^2 = r^2$, $x = r \cos \theta$, and $y = r \sin \theta$, substitution and algebraic simplification yields

$\left(x - \frac{\sqrt{2}}{2}\right)^2 +$ _____ $= 1$. This equation represents _____ with center _____ and radius $r = 1$.

10.6   GRAPHING IN POLAR COORDINATES

OBJECTIVE A :   Given an equation $F(r,\theta) = 0$ in polar coordinates, analyze and sketch its graph.

43.   The graph of $F(r,\theta) = 0$ is symmetric about the x-axis if the equation is unchanged when _____ is replaced by _____.

44.   The graph of $F(r,\theta) = 0$ is symmetric about the origin if the equation is unchanged when r is replaced by _____, or θ is replaced by _____.

45.   The graph of $F(r,\theta) = 0$ is symmetric about the y-axis if the equation is unchanged when θ is replaced by _____, or r is replaced by _____ and θ is replaced by _____.

---

40.   2, $\sin \theta \cos \theta$, $r \cos \theta$, $2xy$, a hyperbola, $C(0,0)$

41. $2r \cos \theta$, $3 + 2x$, $(3 + 2x)^2$, $4x^2$, $3x^2$, -3, $\dfrac{y^2}{3}$

42. $2 \cos \theta \sin \frac{\pi}{4}$, $\sqrt{2} \sin \theta + \sqrt{2} \cos \theta$, $\sqrt{2} r \cos \theta$, $\left(y - \frac{\sqrt{2}}{2}\right)^2$, a circle, $C\left(\frac{\sqrt{2}}{2}, \frac{\sqrt{2}}{2}\right)$

43. θ, $-\theta$ 44. $-r$, $\theta + \pi$ 45. $\pi - \theta$, $-r$, $-\theta$

46. Consider the curve given by
r = 1 - 2 cos θ. Since
cos (-θ) = cos θ, the curve
is symmetric about the _____.

Next, $\frac{dr}{d\theta}$ = _____. Thus,
as θ varies from 0 to $\frac{\pi}{3}$,
r increases from r = _____
to r = _____; and as θ
varies from $\frac{\pi}{2}$ to π, r
increases from r = _____ to
r = _____. Complete the
following table of values for
the curve, and sketch its graph
using its symmetries.

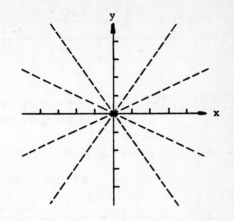

θ	0	π/6	π/3	π/2	2π/3	5π/6	π
r							

47. Consider the curve given by
r^2 = sin θ. Since the sin θ
must be nonnegative in order to
equal the square of a real number,
we must restrict θ to the
interval _____. Since the
equation remains unchanged when r
is replaced by -r, the curve is
symmetric about the _____.
Also, since sin (π - θ) = sin θ,
the curve is symmetric about the
_____. Complete the following
table and sketch the graph using
these symmetries:

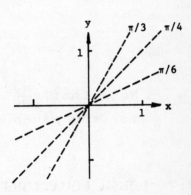

---

46. x-axis, 2 sin θ, -1, 0, 1, 3

θ	0	π/6	π/3	π/2	2π/3	5π/6	π
r	-1	1 - √3̄	0	1	2	1 + √3̄	3

The dashed portion of the curve is the rest
of it due to its symmetry about the x-axis.

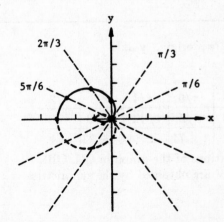

θ	0	$\pi/6$	$\pi/4$	$\pi/3$	$\pi/2$
r^2					
r					

OBJECTIVE B: If $r = f(\theta)$ is differentiable, find the slope $\dfrac{dy}{dx}$ at the point (r, θ) on the graph of f.

48. If $r = f(\theta)$ is differentiable and $\dfrac{dx}{d\theta} \neq 0$, then

slope at (r, θ) = _____.

49. For the polar curve $r = 1 - 2 \cos \theta$ in Problem 46,

$r' =$ _____. When $\theta = \dfrac{2\pi}{3}$ radians, $r =$ _____ and

$r' =$ _____. Thus,

$$\text{Slope at } \left(2, \frac{2\pi}{3}\right) = \frac{\sqrt{3} \, \sin\left(\frac{2\pi}{3}\right) + 2 \cos\left(\frac{2\pi}{3}\right)}{\underline{\hspace{3cm}}}$$

$$= \frac{1/2}{\underline{\hspace{1.5cm}}} = \underline{\hspace{2.5cm}}$$

$$\approx -0.192 \; .$$

Note that $\dfrac{dx}{d\theta} = \dfrac{dr}{d\theta} \cos \theta - r \sin \theta = \dfrac{-3\sqrt{3}}{2} \neq 0$ when $\theta = \dfrac{2\pi}{3}$ for the given polar curve.

10.7   POLAR EQUATIONS OF CONIC SECTIONS

OBJECTIVE A: Given an equation of a straight line in polar coordinates, find a corresponding Cartesian equation.

---

47. $0 \leq \theta \leq \pi$, origin, y-axis

θ	0	$\pi/6$	$\pi/4$	$\pi/3$	$\pi/2$
r^2	0	$1/2$	$1/\sqrt{2}$	$\sqrt{3}/2$	1
r	0	$\pm.71$	$\pm.84$	$\pm.93$	±1

The portions of the graph in QII, QIII, and QIV are obtained by the symmetries.

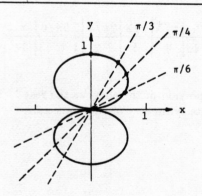

48. $\dfrac{r' \sin \theta + r \cos \theta}{r' \cos \theta - r \sin \theta}$ where $r' = \dfrac{dr}{d\theta}$

49. $2 \sin \theta$, 2, $\sqrt{3}$, $\sqrt{3} \cos\left(\frac{2\pi}{3}\right) - 2 \sin\left(\frac{2\pi}{3}\right)$, $\dfrac{-3\sqrt{3}}{2}$, $\dfrac{-1}{3\sqrt{3}}$

50. If $P_0(r_0, \theta_0)$ is the foot of the perpendicular from the origin
     to the line L with $r_0 \geq 0$, then an equation for L in
     polar coordinates is

     _____ .

51. A polar equation for the line
     displayed in the figure at the
     right is

     _____ .

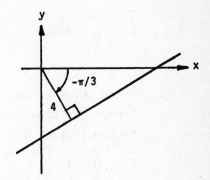

52. Using the identity for $\cos(A + B)$, we find the following
     Cartesian equation for the line in Problem 51:

$$r \cos\left(\theta + \frac{\pi}{3}\right) = 4$$

$$r\left[\underline{\hspace{4cm}}\right] = 4$$

$$\frac{1}{2} r \cos\theta - \underline{\hspace{2cm}} = 4$$

$$\underline{\hspace{4cm}} = 4$$

     or

$$y = \underline{\hspace{3cm}} .$$

OBJECTIVE B : Given the eccentricity of an ellipse together with its
                semimajor axis, write a polar equation for the ellipse.

53. The semimajor axis for the orbit of Saturn is 9.539 AU
     (astronomical units) and the eccentricity is 0.0543. Thus a
     polar equation of Saturn's orbit around the sun is

$$r = \frac{\underline{\hspace{1cm}}\left[1 - (0.0543)^2\right]}{\underline{\hspace{4cm}}}$$

$$= \frac{9.511}{1 + 0.0543 \cos\theta}$$

$$= \frac{\underline{\hspace{2cm}}}{18.42 + \cos\theta} .$$

     At its most distant point (aphelion), Saturn is _____
     (astronomical units) from the sun.

---

50. $r \cos(\theta - \theta_0) = r_0$

51. $r \cos\left(\theta + \frac{\pi}{3}\right) = 4$

52. $\cos\theta \cos\frac{\pi}{3} - \sin\theta \sin\frac{\pi}{3}$, $\frac{\sqrt{3}}{2} r \sin\theta$, $\frac{1}{2} x - \frac{\sqrt{3}}{2} y$, $\frac{1}{\sqrt{3}}(x - 8)$

53. $\frac{9.539\left[1 - (0.0543)^2\right]}{1 + 0.0543 \cos\theta}$, 175.2, 10 AU

10.8   INTEGRATION IN POLAR COORDINATES

[OBJECTIVE A]:   Find the total plane area enclosed by a polar graph
$r = f(\theta)$ and the rays $\theta = \alpha$, $\theta = \beta$.

54.   The area bounded by the polar curve $r = f(\theta)$ and the rays
$\theta = \alpha$, $\theta = \beta$ is given by the integral

$$A = \underline{\hspace{4cm}}.$$

55.   Find the area inside the larger loop and outside the smaller
loop of the polar graph $r = 1 - 2 \cos \theta$ given in Problem 46.
<u>Solution</u>.   The graph of the curve is sketched in the figure
below.   That part of the curve traced out as θ varies from
$\theta = 0$ to $\theta = \pi$ is drawn in with a broader ink stroke.   Now,
as θ varies from $\frac{\pi}{3}$ to
π,   the radius vector r
sweeps out the larger loop of
the curve including that portion
of the smaller loop lying above
the x-axis.   By symmetry, the
area of that smaller half-loop
is the same as the area of the
half-loop swept out by r as
θ varies from 0 to $\frac{\pi}{3}$.
Thus, the total area inside the
larger loop and outside the
smaller loop is

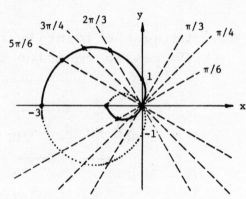

$$A = 2 \left[\int_{\pi/3}^{\pi} \tfrac{1}{2} f^2(\theta) \; d\theta - \underline{\hspace{4cm}} \right]$$

$$= \int_{\pi/3}^{\pi} (1 - 4 \cos \theta + 4 \cos^2 \theta) \; d\theta - \underline{\hspace{5cm}}$$

$$= [\theta - 4 \sin \theta + 4(\tfrac{\theta}{2} + \tfrac{1}{4} \sin 2\theta]_{\pi/3}^{\pi} - \underline{\hspace{4cm}}$$

$$= (\pi + 2\pi) - [\tfrac{\pi}{3} - 4\left(\tfrac{\sqrt{3}}{2}\right) + 4\left(\tfrac{\pi}{6} + \tfrac{1}{4} \cdot \tfrac{\sqrt{3}}{2}\right)] - \underline{\hspace{4cm}}$$

$$= \underline{\hspace{3cm}} \approx 8.338.$$

---

54. $\int_{\alpha}^{\beta} \tfrac{1}{2} \left[f(\theta) \right]^2 \; d\theta$

55. $\int_{0}^{\pi/3} \tfrac{1}{2} f^2(\theta) \; d\theta$, $\int_{0}^{\pi/3} (1 - 4 \cos \theta + 4 \cos^2 \theta) \; d\theta$, $[\theta - 4 \sin \theta + 4\left(\tfrac{\theta}{2} + \tfrac{1}{4} \sin 2\theta\right]_{0}^{\pi/3}$,

$[\tfrac{\pi}{3} - 4\left(\tfrac{\sqrt{3}}{2}\right) + 4\left(\tfrac{\pi}{6} + \tfrac{1}{4} \cdot \tfrac{\sqrt{3}}{2}\right)]$, $3\sqrt{3} + \pi$

OBJECTIVE B :   Given a polar curve $r = f(\theta)$,   calculate its arc
length as θ varies from $\theta = a$ to $\theta = b$.

56. The differential element of arc length   ds   for the polar curve
$r = f(\theta)$ satisfies the equation

$$ds^2 = \underline{\hspace{3cm}}.$$

Thus the length of arc traced out by the curve as θ varies
from $\theta = a$ to $\theta = b$ is given by

$$s = \underline{\hspace{4cm}}.$$

57. To determine the length of the curve $r = 3 \sec \theta$ as θ varies
from $\theta = 0$ to $\theta = \frac{\pi}{4}$,   we find $\frac{dr}{d\theta} = \underline{\hspace{2cm}}.$ Then the
arc length is,

$$s = \int_0^{\pi/4} \sqrt{\underline{\hspace{3cm}}}\, d\theta = 3 \int_0^{\pi/4} \sec\theta \sqrt{\underline{\hspace{2cm}}}\, d\theta$$

$$= 3 \int_0^{\pi/4} \underline{\hspace{2cm}}\, d\theta = \underline{\hspace{2cm}}\Big]_0^{\pi/4} = \underline{\hspace{2cm}}.$$

58. Consider the polar curve

$r = \cos^4 \frac{\theta}{4}$.   As θ varies
from $\theta = 0$ to $\theta = 2\pi$,   the
equation describes the path
shown in the figure at the
right.   As θ varies from
$\theta = 2\pi$ to $\theta = 4\pi$ the curve
shown is reflected across the
x-axis.   Thus, the total arc
length is given by

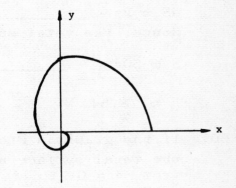

$$s = 2 \int_0^{2\pi} \sqrt{r^2 + \underline{\hspace{1.5cm}}}\, d\theta.$$

Now, $\frac{dr}{d\theta} = \underline{\hspace{2cm}}$ so that

$$r^2 + \left(\frac{dr}{d\theta}\right)^2 = \cos^8 \frac{\theta}{4} + \underline{\hspace{2cm}} = \cos^6 \frac{\theta}{4}.$$

Thus, $s = 2 \int_0^{2\pi} \underline{\hspace{2.5cm}}\, d\theta.$

Since $\cos \frac{\theta}{4} \geq 0$ for $0 \leq \theta \leq 2\pi$,   the integral becomes

---

56. $r^2\, d\theta^2 + dr^2$, $\int_a^b \sqrt{r^2 + \left(\frac{dr}{d\theta}\right)^2}\, d\theta$

57. $3 \sec\theta \tan\theta$, $9 \sec^2\theta + 9 \sec^2\theta \tan^2\theta$, $1 + \tan^2\theta$, $\sec^2\theta$, $3 \tan\theta$,   3

$$s = 2 \int_0^{2\pi} (1 - \sin^2 \tfrac{\theta}{4}) \underline{\hspace{3cm}} d\theta$$

$$= 2 \int \underline{\underline{\hspace{0.5cm}}} \; \underline{\hspace{2.5cm}} \; du, \quad \text{where} \quad u = \sin \tfrac{\theta}{4}$$

$$= \underline{\hspace{1.5cm}} \; (\underline{\hspace{1.5cm}})] \underline{\underline{\hspace{0.5cm}}} = \underline{\hspace{2cm}}.$$

OBJECTIVE C : Find the area of the surface generated when a polar
graph is revolved about the x-axis or the y-axis.

59. The graph of the polar equation
$r = 5 \cos \theta$ is the circle shown
at the right. If the graph is
rotated about the x-axis, the
total surface area is generated
by that portion of the graph as
θ varies from $\theta = 0$ to
$\theta = \underline{\hspace{2cm}}$ because of the
symmetry of the graph across the x-axis.

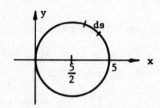

An element of arc length ds (see the figure) generates a
portion of surface area dS = $\underline{\hspace{2cm}}$ ds, where $y = r \sin \theta$
and ds = $\underline{\hspace{2cm}}$. Thus,

$$dS = 2\pi \cdot \underline{\hspace{6cm}} d\theta = 2\pi(5 \cos \theta) \underline{\hspace{2cm}} d\theta.$$

Hence, the total surface area is given by

$$S = 50\pi \int \underline{\underline{\hspace{0.5cm}}} \; \underline{\hspace{2.5cm}} \; d\theta = \underline{\hspace{4cm}}] \underline{\underline{\hspace{0.5cm}}} = \underline{\hspace{2cm}}$$

$$\approx 78.54.$$

60. If the graph in Problem 59 is rotated about the axis $\theta = \tfrac{\pi}{2}$,
the total surface area is generated by the graph as θ varies
from $\theta = 0$ to $\theta = \underline{\hspace{2cm}}$. An element of arc length ds
now generates a portion of the surface area

$$dS = \underline{\hspace{3cm}} = 2\pi \left(\underline{\hspace{2cm}}\right) \cdot 5 \; d\theta.$$

Thus, the total surface area is given by

$$S = 50\pi \int \underline{\underline{\hspace{0.5cm}}} \; \underline{\hspace{2.5cm}} \; d\theta = 50\pi \; [\underline{\hspace{4cm}}] \underline{\underline{\hspace{0.5cm}}}$$

$$= \underline{\hspace{3cm}} \approx 246.74.$$

---

58. $\left(\dfrac{dr}{d\theta}\right)^2$, $-\cos^3 \tfrac{\theta}{4} \sin \tfrac{\theta}{4}$, $\cos^6 \tfrac{\theta}{4} \sin^2 \tfrac{\theta}{4}$, $\left|\cos^3 \tfrac{\theta}{4}\right|$, $\cos \tfrac{\theta}{4}$, $\int_0^1 4(1 - u^2) \, du$, $8(u - \tfrac{1}{3} u^3)]_0^1$, $\tfrac{16}{3}$

59. $\tfrac{\pi}{2}$, $2\pi y$, $\sqrt{dr^2 + r^2 \, d\theta^2}$, $r \sin \theta \sqrt{25 \sin^2 \theta + 25 \cos^2 \theta}$, $5 \sin \theta$,

$\int_0^{\pi/2} \sin \theta \cos \theta \, d\theta$, $50\pi \cdot \tfrac{1}{2} \sin^2 \theta]_0^{\pi/2}$, 25π

60. π, $2\pi x \, ds$, $r \cos \theta$, $\int_0^\pi \cos^2 \theta \, d\theta$, $\tfrac{1}{2} \theta + \tfrac{1}{4} \sin 2\theta]_0^\pi$, $25\pi^2$

CHAPTER 10   SELF-TEST

1. Find the eccentricity of the ellipse $81x^2 + 25y^2 = 2025$.

2. Consider the equation $x^2 - 2xy - y^2 = 12$.
    (a) Use the discriminant to classify it.
    (b) Transform the equation by a rotation of axes into an equation with no cross-product term.

3. Write an equation for the circle with center $C(4,6)$ and radius $r = 3$.

4. Sketch the graph of the curve described by $x = t - 2$ and $y = t^2 - t + 1$ for $-\infty < t < \infty$. Also find a Cartesian equation of the curve.

5. Find parametric equations for the circle $x^2 + y^2 = 2x$, using as parameter the arc length s measured counterclockwise from the point $(2,0)$ to the point (x,y) .

6. Given the parametric equations $x = t^2 - 1$ and $y = t + 1$
    (a) Express dx and dy in terms of t and dt,
    (b) Find $\dfrac{d^2y}{dx^2}$ in terms of t,
    (c) Find an equation of the tangent line to the curve at the point for which $t = 1$.

7. Consider the curve given by the parametric equations $x = 3t - 1$ and $y = t^2 - t$. Eliminate the parameter t to find an equation of the form $y = F(x)$ and find $\dfrac{dy}{dx}$ in terms of $\dfrac{dy}{dt}$ and $\dfrac{dx}{dt}$. For what value of t is $\dfrac{dy}{dx} = 0$? Sketch the graph of the curve over the interval $-2 \leq t \leq 3$.

8. Find the length of the curve given by $x = t^3 + 3t^2$ and $y = t^3 - 3t^2$ for $0 \leq t \leq 2$.

9. Find the area of the surface generated by rotating the cardiod $x = 2 \cos \theta - \cos 2\theta$, $y = 2 \sin \theta - \sin 2\theta$, for $0 \leq \theta \leq \pi$, about the x-axis. First sketch the curve.

10. Convert the following from polar coordinates to Cartesian coordinates.

     (a) $\left(-6, \dfrac{\pi}{4}\right)$ (b) $\left(1, -\dfrac{5\pi}{6}\right)$ (c) $\left(-2, -\dfrac{7\pi}{12}\right)$

11. Write the following simple polar equations in Cartesian form.

     (a) $r = 5$ (b) $\theta = \dfrac{3\pi}{4}$ (c) $r = -5 \csc \theta$

In Problems 12 and 13, graph the polar equation.

12. $r = 2 \cos 4\theta$ 13. $r^2 = -\sin 2\theta$

14. Determine a Cartesian equation, and sketch the curve, for $r = \cos\theta + 5\sin\theta$.

15. Find a polar equation of the line with slope $m = -2$ passing through the Cartesian point $(1,-3)$.

16. Find a polar equation of the circle centered at the Cartesian point $(\frac{1}{3},0)$ and passing through the origin.

17. Find the length of the polar curve $r = a\cos(\theta + b)$ from $\theta = 0$ to $\theta = \pi$, where a and b are constants.

18. Find the area of the region bounded on the outside by the graph of $r = 2 + 2\sin\theta$ for $\theta = 0$ to $\theta = \pi$, and on the inside by the graph of $r = 2\sin\theta$.

19. Write an integral expressing the surface area generated by rotating the portion of the polar curve $r = 1 + \cos\theta$ in the first quadrant about $\theta = \frac{\pi}{2}$.

SOLUTIONS TO CHAPTER 10 SELF-TEST

1. The standard form of the ellipse is

$$\frac{x^2}{25} + \frac{y^2}{81} = 1 \ .$$

Thus, since $81 > 25$, the foci lie along the y-axis at $(0, \pm c)$ where $c = \sqrt{81 - 25} = 2\sqrt{14}$. The eccentricity is $e = \frac{c}{a} = \frac{2\sqrt{14}}{9} \approx 0.83$.

2. (a) The discriminant is $B^2 - 4AC = (-2)^2 - 4(1)(-1) = 8 > 0$. Therefore, the equation represents a hyperbola.

   (b) $\cot 2\alpha = \frac{A - C}{B} = \frac{1 + 1}{-2} = -1$; thus $2\alpha = \frac{3\pi}{4}$ radians.

   Hence, $\cos 2\alpha = \frac{-1}{\sqrt{2}}$. Using the half-angle formulas,

   $$\sin \alpha = \sqrt{\frac{1 - \cos 2\alpha}{2}} = \sqrt{\frac{1 + \sqrt{2}}{2\sqrt{2}}} = \frac{1}{2}\sqrt{2 + \sqrt{2}}, \quad \text{and}$$

   $$\cos \alpha = \sqrt{\frac{1 + \cos 2\alpha}{2}} = \sqrt{\frac{1 - \sqrt{2}}{2\sqrt{2}}} = \frac{1}{2}\sqrt{2 - \sqrt{2}}.$$

   Now, $A' = A \cos^2 \alpha + B \cos \alpha \sin \alpha + C \sin^2 \alpha$
   $$= \frac{1}{4}(2 - \sqrt{2}) - \frac{2}{4}(\sqrt{-2 + 4}) - \frac{1}{4}(2 + \sqrt{2}) = - \sqrt{2}$$

   $C' = A \sin^2 \alpha - B \cos \alpha \sin \alpha + C \cos^2 \alpha$
   $$= \frac{1}{4}(2 + \sqrt{2}) + \frac{2}{4}(\sqrt{-2 + 4}) - \frac{1}{4}(2 - \sqrt{2}) = \sqrt{2}.$$

   Thus, since $B' = 0$ because of the choice $\alpha = \frac{3\pi}{8}$, the original equation is reduced to $A'x'^2 + C'y'^2 = F$, or $-\sqrt{2}x'^2 + \sqrt{2}y'^2 = 12$, or $y'^2 - x'^2 = 6\sqrt{2}$.

3. $(x - 4)^2 + (y - 6)^2 = 9$, or $x^2 + y^2 - 8x - 12y + 43 = 0$.

4. We have the following table giving some of the points on the curve:

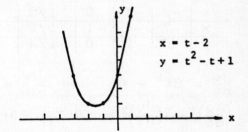

$x = t - 2$
$y = t^2 - t + 1$

t	-1	0	1	2	3
x	-3	-2	-1	0	1
y	3	1	1	3	7

Substitution of $t = x + 2$ into the parametric equation for y gives

$y = (x + 2)^2 - (x + 2) + 1$, or simplifying algebraically,

$y = x^2 + 3x + 3$. This is a Cartesian equation of the parabola sketched in the figure above.

5. Completing the square gives the
    equation $(x - 1)^2 + y^2 = 1$
    which we recognize as an equation
    of a unit circle centered at
    $(1,0)$ (see the figure at the
    right). Hence, $x - 1 = 1 \cos s$
    and $y - 0 = 1 \sin s$ or,

    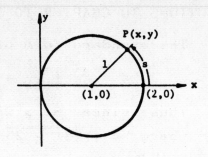

$$x = 1 + \cos s \quad \text{and} \quad y = \sin s$$

   are parametric equations for
   the circle in terms of the arc
   length parameter s.

6. (a) $\frac{dx}{dt} = 2t$ and $\frac{dy}{dt} = 1$, so that $dx = 2t\,dt$ and $dy = dt$.

    (b) $\frac{dy}{dx} = \frac{1}{2t}$ so that $\frac{d^2y}{dx^2} = \frac{dy'/dt}{dx/dt} = \frac{-1/2t^2}{2t} = -\frac{1}{4t^3}$

    (c) When $t = 1$, $x = 0$, $y = 2$, and $\frac{dy}{dx} = \frac{1}{2}$; thus
        $y - 2 = \frac{1}{2}(x - 0)$ or $2y - x = 4$ is an equation of the
        tangent line.

7. $\frac{dx}{dt} = 3$ and $\frac{dy}{dt} = 2t - 1$, so that $\frac{dy}{dx} = \frac{dy/dt}{dx/dt} = \frac{1}{3}(2t - 1)$. When
    $t = \frac{1}{2}$, $\frac{dy}{dx} = 0$ which occurs at the point $(x,y) = \left(\frac{1}{2}, -\frac{1}{4}\right)$. To
    eliminate the parameter t, use the parametric expression for
    x, $t = \frac{1}{3}(x + 1)$, and substitute into the parametric
    expression for y giving

    $y = \frac{1}{9}(x + 1)^2 - \frac{1}{3}(x + 1)$

    $= \frac{1}{9}(x^2 - x - 2)$.

    A table of coordinate values
    is as follows:

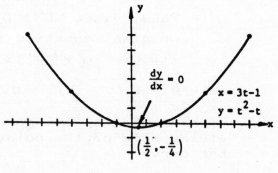

t	-2	-1	0	1/2	1	2	3
x	-7	-4	-1	1/2	2	5	8
y	6	2	0	-1/4	0	2	6

8. $\frac{dx}{dt} = 3t^2 + 6t$ and $\frac{dy}{dt} = 3t^2 - 6t$ so that

$$\left(\frac{dx}{dt}\right)^2 + \left(\frac{dy}{dt}\right)^2 = (9t^4 + 36t^3 + 36t^2) + (9t^4 - 36t^3 + 36t^2)$$

$$= 18t^2(t^2 + 4).$$

Thus, the arc length is given by

$$s = \int_0^2 3\sqrt{2}\, t\sqrt{t^2 + 4}\; dt = 3\sqrt{2}\left(\frac{1}{3}\right)\left(t^2 + 4\right)^{3/2}\Big]_0^2$$

$$= 8(4 - \sqrt{2}) \approx 20.7 \text{ units}.$$

9. The required surface is obtained by rotating the arc from $\theta = 0$ to $\theta = \pi$ about the x-axis. The arc is shown in the figure at the right.

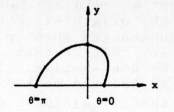

$$\frac{dx}{d\theta} = -2 \sin\theta + 2 \sin 2\theta,$$

$$\frac{dy}{d\theta} = 2 \cos\theta - 2 \cos 2\theta \quad \text{so that}$$

$$\left(\frac{dx}{d\theta}\right)^2 + \left(\frac{dy}{d\theta}\right)^2 = \left(4 \sin^2\theta - 8 \sin\theta \sin 2\theta + 4 \sin^2 2\theta\right) +$$
$$\left(4 \cos^2\theta - 8 \cos\theta \cos 2\theta + 4 \cos^2 2\theta\right)$$
$$= 8(1 - \sin\theta \sin 2\theta - \cos\theta \cos 2\theta)$$
$$= 8[1 - \cos(2\theta - \theta)]$$
$$= 8(1 - \cos\theta)$$

Therefore, the surface area is given by

$$S = \int_0^\pi 2\pi(2 \sin\theta - \sin 2\theta) \cdot 2\sqrt{2}\ \sqrt{1 - \cos\theta}\ d\theta$$

$$= \int_0^\pi 8\sqrt{2}\ \pi \sin\theta(1 - \cos\theta)^{3/2}\ d\theta, \quad \text{with} \quad \sin 2\theta = 2 \sin\theta \cos\theta$$

$$= \frac{16\sqrt{2}}{5}\ \pi(1 - \cos\theta)^{5/2}\Big]_0^\pi = \frac{128\pi}{5} \approx 80.4 \quad \text{square units.}$$

10. (a) $x = -6 \cos\frac{\pi}{4} = -6 \cdot \frac{\sqrt{2}}{2} = -3\sqrt{2};\quad y = -6 \sin\frac{\pi}{4} = -3\sqrt{2}$

(b) $x = 1 \cos\left(-\frac{5\pi}{6}\right) = \cos\frac{5\pi}{6} = -\frac{\sqrt{3}}{2};$

$y = 1 \sin\left(-\frac{5\pi}{6}\right) = -\sin\frac{5\pi}{6} = -\frac{1}{2}$

(c) $x = -2 \cos\left(-\frac{7\pi}{12}\right) = -2 \cos\frac{7\pi}{12} = \frac{\sqrt{2}}{2}\ (\sqrt{3} - 1) \approx 0.518$

$y = -2 \sin\left(-\frac{7\pi}{12}\right) = 2 \sin\frac{7\pi}{12} = \frac{\sqrt{2}}{2}\ (\sqrt{3} + 1) \approx 1.932$

11. (a) $\pm\sqrt{x^2 + y^2} = 5\quad\text{or}\quad x^2 + y^2 = 25$

(b) $y = -x$

(c) Equivalently, $r \sin\theta = -5$, or $y = -5$

12. $r = 2 \cos 4\theta$ is symmetric about the x-axis, the y-axis, and the origin.

The graph is the eight-leafed rose sketched at the right.

13. $r^2 = -\sin 2\theta$ is symmetric about
the origin since the equation is
unchanged when r is replaced by
$-r$. Notice that $-\sin 2\theta$ must
be nonnegative. If θ is
restricted to the interval $[0, 2\pi]$,
then $-\sin 2\theta \geq 0$ if and only
if θ is in $[\frac{\pi}{2}, \pi]$ or $[\frac{3\pi}{2}, 2\pi]$.
Using the following table and
symmetry we obtain the graph
sketched at the right:

θ	$\pi/2$	$7\pi/12$	$3\pi/4$	π	$3\pi/2$
r	0	$\sqrt{2}/2$	1	0	0

The curve is a lemniscate.

14.   Since $r = 5\cos\theta + 5\sin\theta$, for
$r \neq 0$ ($r = 0$ is on the graph at
$\theta = \frac{3\pi}{4}$), $r^2 = 5r\cos\theta + 5r\sin\theta$,
or $x^2 + y^2 = 5x + 5y$. Then,
completing the squares in the x
and y terms gives

$$\left(x - \frac{5}{2}\right)^2 + \left(y - \frac{5}{2}\right)^2 = \frac{25}{2}.$$

This is a circle with center
$\left(\frac{5}{2}, \frac{5}{2}\right)$ and radius $\frac{5}{\sqrt{2}}$.

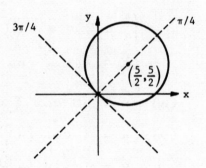

15.   A Cartesian equation of the line is given by $y + 3 = -2(x - 1)$
or $2x + y = -1$. Thus, $2r\cos\theta + r\sin\theta = -1$, or solving
for r, $r = \dfrac{-1}{2\cos\theta + \sin\theta}$.

16.   A Cartesian equation of the circle is given by $\left(x - \frac{1}{3}\right)^2 + y^2 = \frac{1}{9}$
or $x^2 + y^2 = \frac{2}{3}x$. Thus, $3r^2 = 2r\cos\theta$, or since $r = 0$ lies
on the graph $3r = 2\cos\theta$ when $\theta = \frac{\pi}{2}$, the latter gives a polar
equation of the circle.

17.   For $r = a\cos(\theta + b)$, $\frac{dr}{d\theta} = -a\sin(\theta + b)$ so that
$r^2 + \left(\frac{dr}{d\theta}\right)^2 = a^2\cos^2(\theta + b) + a^2\sin^2(\theta + b) = a^2$. Therefore
the arc length is given by the integral $s = \displaystyle\int_0^\pi \sqrt{a^2}\, d\theta = |a|\pi$.

18.   A graph depicting the region is
       shown in the figure at the right
       (the shaded portion represents
       the area we seek). Thus the
       area is given by

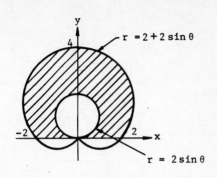

$$A = \frac{1}{2} \int_0^\pi (2 + 2 \sin \theta)^2 \, d\theta$$

$$- \frac{1}{2} \int_0^\pi (2 \sin \theta)^2 \, d\theta$$

$$= \frac{1}{2} \int_0^\pi (4 + 8 \sin \theta) \, d\theta$$

$$= 2(\theta - 2 \cos \theta)]_0^\pi$$

$$= 2\pi + 8 \approx 14.28.$$

19.   A sketch of the surface is shown
       in the figure at the right. An
       element of arc length ds
       generates a portion of surface
       area

$$dS = 2\pi x \, dx.$$

Now, $\dfrac{ds}{d\theta} = \sqrt{r^2 + \left(\dfrac{dr}{d\theta}\right)^2}$

$$= \sqrt{1 + 2 \cos \theta + \cos^2 \theta) + \sin^2 \theta} = \sqrt{2} \, \sqrt{1 + \cos \theta}$$

Hence, $dS = 2\pi x \, ds = 2\pi r \cos \theta \, ds = 2\sqrt{2} \pi (1 + \cos \theta)^{3/2} \cos \theta \, d\theta.$
Therefore, the total surface area generated is given by the
integral

$$S = 2\sqrt{2} \pi \int_0^{\pi/2} (1 + \cos \theta)^{3/2} \cos \theta \, d\theta.$$

(The definite integral can be evaluated by using the identity
$\cos^2 \frac{\theta}{2} = 1 + \cos \theta,$ but it is tedious to carry out the
calculations. Using Simpson's rule with n = 12, an
approximate value to the integral is 39.31. This calculation
was made using a calculator.)

NOTES:

CHAPTER 11 VECTORS AND ANALYTIC GEOMETRY IN SPACE

11.1   VECTORS IN THE PLANE

OBJECTIVE A : Given the points P_1 and P_2 in the plane, express the vector $\overrightarrow{P_1P_2}$ in the form $a\vec{i} + b\vec{j}$.

1. Consider the points $P_1(x_1, y_1)$ and $P_2(x_2, y_2)$. The directed change in the x-direction from P_1 to P_2 is $a = $ _____, and in the y-direction it is $b = $ _____. Hence, the vector $\overrightarrow{P_1P_2}$ can be expressed as $\overrightarrow{P_1P_2} = $ _____.

2. The vector from the point $P_1(-7, 4)$ to the point $P_2(3, -5)$ is $\overrightarrow{P_1P_2} = (3 - \text{\_\_})\vec{i} + (\text{\_\_} - 4)\vec{j}$ or $\overrightarrow{P_1P_2} = $ _____.

3. The vector from $P(2, -9)$ to the origin is $\overrightarrow{PO} = $ _____.

OBJECTIVE B : Express the sum and difference of two given vectors, and multiples of given vectors by scalars, in the form $a\vec{i} + b\vec{j}$.

4. The sum of $\vec{v}_1 = a_1\vec{i} + b_1\vec{j}$ and $\vec{v}_2 = a_2\vec{i} + b_2\vec{j}$ is $\vec{v}_1 + \vec{v}_2 = $ _____.

5. If $\vec{v}_1 = 3\vec{i} - 4\vec{j}$ and $\vec{v}_2 = -5\vec{i} + 2\vec{j}$, then $\vec{v}_1 + \vec{v}_2 = $ _____ and $\vec{v}_2 - \vec{v}_1 = $ _____.

6. If $\vec{v}_1 = -3\vec{i} + 7\vec{j}$, then
$-\vec{v} = $ _____, $2\vec{v} = $ _____,
$-5\vec{v} = $ _____, and $\frac{1}{4}\vec{v} = $ _____.

OBJECTIVE C : Given a vector $\vec{v} = a\vec{i} + b\vec{j}$, calculate its length or magnitude, direction, and the angle it makes with the positive x-axis.

7. The length or magnitude of the vector $\vec{v} = a\vec{i} + b\vec{j}$ is given by $|\vec{v}| = $ _____.

---

1. $x_2 - x_1$, $y_2 - y_1$, $(x_2 - x_1)\vec{i} + (y_2 - y_1)\vec{j}$ 2. -7, -5, $10\vec{i} - 9\vec{j}$ 3. $-2\vec{i} + 9\vec{j}$

4. $(a_1 + a_2)\vec{i} + (b_1 + b_2)\vec{j}$, $(a_2 - a_1)\vec{i} + (b_2 - b_1)\vec{j}$ 5. $-2\vec{i} - 2\vec{j}$, $-8\vec{i} + 6\vec{j}$

6. $3\vec{i} - 7\vec{j}$, $-6\vec{i} + 14\vec{j}$, $15\vec{i} - 35\vec{j}$, $-\frac{3}{4}\vec{i} + \frac{7}{4}\vec{j}$ 7. $\sqrt{a^2 + b^2}$

8. The angle θ that $\vec{v} = a\vec{i} + b\vec{j}$ makes with the positive x-axis satisfies

$$\cos\theta = \underline{\hspace{2cm}} \quad \text{and} \quad \sin\theta = \underline{\hspace{2cm}} \; .$$

9. The length of $\vec{v} = -\sqrt{2}\vec{i} + 7\vec{j}$ is $|\vec{v}| = \underline{\hspace{4cm}}$.

10. Consider the vector $\vec{v} = -\vec{i} + \sqrt{3}\vec{j}$. The length of $\vec{v}$ is $|\vec{v}| = \underline{\hspace{2cm}}$. The direction of $\vec{v}$ is given by Direction of $\vec{v} = \frac{\vec{v}}{|\vec{v}|} = \underline{\hspace{3cm}}$. Thus, if θ is the angle $\vec{v}$ makes with the positive x-axis, then

$$\cos\theta = \underline{\hspace{2cm}} \quad \text{and} \quad \sin\theta = \underline{\hspace{2cm}} \; .$$

It follows that $\theta = \frac{\pi}{2} + \underline{\hspace{2cm}} = \underline{\hspace{2cm}}$ radians.

OBJECTIVE D : Find unit vectors tangent and normal to a given curve $y = f(x)$ at a specified point $P(a,b)$.

11. Consider the curve $y = 1 - 2e^x$. Observe that the point $P(0,-1)$ lies on the curve. The slope of the line tangent to curve at P is $y' = \underline{\hspace{2cm}}|_{x=0} = \underline{\hspace{2cm}}$. A vector $\vec{v}$ having this slope is given by $\vec{v} = \underline{\hspace{2cm}}\vec{i} + \underline{\hspace{2cm}}\vec{j}$. (Any nonzero multiple of $\vec{v}$ will also have the slope -2, so $\vec{v}$ is not unique.) A unit vector parallel to $\vec{v}$ is given by

$$\vec{u} = \frac{\vec{v}}{|\vec{v}|} = \underline{\hspace{3cm}} \; .$$

The vector $\vec{u}$ is a unit vector tangent to the curve at $P(0,-1)$; the vector $-\vec{u}$ is also a unit vector tangent to the curve at $P(0,-1)$.

12. One unit vector normal to the curve at P for Problem 11 above is $\vec{n} = \underline{\hspace{3cm}}$. A second vector is $-\vec{n} = \underline{\hspace{3cm}}$. Either vector will do as a unit vector normal to $y = 1 - 2e^x$ at $P(0,-1)$.

11.2 CARTESIAN (RECTANGULAR) COORDINATES AND VECTORS IN SPACE

OBJECTIVE A : Given two points P_1 and P_2 in space, express the vector $\vec{P_1P_2}$ from P_1 to P_2 in the form $a\vec{i} + b\vec{j} + c\vec{k}$.

13. The vector from the point $(-1,3,0)$ to the point $(4,-2,1)$ is given by $(4 + 1)\vec{i} + \underline{\hspace{2cm}}\vec{j} + \underline{\hspace{2cm}}\vec{k}$ or $\underline{\hspace{2cm}}$.

---

8. $\dfrac{a}{\sqrt{a^2 + b^2}}, \dfrac{b}{\sqrt{a^2 + b^2}}$ 9. $\sqrt{2 + 49} = \sqrt{51}$ 10. $\sqrt{1 + 3} = 2$, $-\frac{1}{2}\vec{i} + \frac{\sqrt{3}}{2}\vec{j}$, $-\frac{1}{2}$, $\frac{\sqrt{3}}{2}$, $\frac{\pi}{6}$, $\frac{2\pi}{3}$

11. $-2e^x$, -2, 1, -2, $\frac{1}{\sqrt{5}}\vec{i} - \frac{2}{\sqrt{5}}\vec{j}$ 12. $\frac{2}{\sqrt{5}}\vec{i} + \frac{1}{\sqrt{5}}\vec{j}$, $-\frac{2}{\sqrt{5}}\vec{i} - \frac{1}{\sqrt{5}}\vec{j}$ 13. $(-2 - 3)$, $(1 - 0)$, $5\vec{i} - 5\vec{j} + \vec{k}$

OBJECTIVE B: Find the length of any space vector.

14. The length of the vector $\vec{v} = -2\vec{i} + 6\vec{j} - 5\vec{k}$ is

$$|\vec{v}| = \sqrt{(-2)^2 + (\underline{\quad}) + (\underline{\quad})} = \sqrt{\underline{\quad}} \approx 8.06 .$$

OBJECTIVE C: Given a Cartesian equation of a sphere in space, find the coordinates of its center and the radius.

15. Consider the equation of the sphere given by

$$x^2 + y^2 + z^2 + 2x - 6y + 4z + 9 = 0.$$

Complete the squares in the given equation to obtain,

$(x^2 + 2x + 1) + (y^2 - 6y + 9) + (\underline{\qquad\qquad}) = \underline{\qquad}$, or

$(x + 1)^2 + (y - 3)^2 + (\underline{\qquad}) = \underline{\quad}$.

Thus, the center is _____ and the radius is _____ .

OBJECTIVE D: Given a nonzero space vector find its direction.

16. If $\vec{A}$ is a nonzero vector, the direction of $\vec{A}$ is a _____ vector obtained from $\vec{A}$ by dividing $\vec{A}$ by _____ .

17. To compute the direction of the vector $\vec{A} = -2\vec{i} + 6\vec{j} - 5\vec{k}$, we divide $\vec{A}$ by $|\vec{A}| = $ _____ found in Problem 14. This gives the unit vector $\vec{u} = \dfrac{\vec{A}}{|\vec{A}|} = $ _____ as the direction of $\vec{A}$.

11.3 DOT PRODUCTS

OBJECTIVE A: Find the scalar product or dot product of two vectors in space and the cosine of the angle between them.

18. The scalar or dot product of the vectors $\vec{A} = 2\vec{i} + \vec{j} - 3\vec{k}$ and $\vec{B} = 3\vec{i} - 6\vec{j} + 2\vec{k}$ is $\vec{A}\cdot\vec{B} = (2)(3) + (1)(\underline{\quad}) + (\underline{\quad})(\underline{\quad})$ = _____ .

19. For the vectors in Problem 18,
$|\vec{A}| = \sqrt{4 + 1 + \underline{\quad}} = \underline{\quad}$ and $|\vec{B}| = \sqrt{9 + 36 + \underline{\quad}} = \underline{\quad}$.
Thus, the cosine of the angle θ between them is

$$\cos\theta = \frac{\underline{\quad}}{|\vec{A}||\vec{B}|} = \frac{-6}{\underline{\quad}} \approx -.229, \quad \text{or} \quad \theta \approx 103°.$$

---

14. 6^2, $(-5)^2$, 65

15. $z^2 + 4z + 4$, $1 + 4$, $(z + 2)^2$, 5, $(-1,3,-2)$, $\sqrt{5}$

16. unit, $|\vec{A}|$

17. $\sqrt{65}$, $-\dfrac{2}{\sqrt{65}}\vec{i} + \dfrac{6}{\sqrt{65}}\vec{j} - \dfrac{5}{\sqrt{65}}\vec{k}$

18. -6, -3, 2, -6

19. 9, $\sqrt{14}$, 4, 7, $\vec{A} \cdot \vec{B}$, $7\sqrt{14}$

OBJECTIVE B: Given two space vectors $\vec{A}$ and $\vec{B}$ find the projection vector $proj_{\vec{A}} \vec{B}$ and the scalar component of $\vec{B}$ in the direction $\vec{A}$.

20.   The projection of $\vec{B}$ onto $\vec{A}$ is the vector
$$proj_{\vec{A}} \vec{B} = \underline{\hspace{2cm}}.$$

21.   The scalar component of $\vec{B}$ in the direction $\vec{A}$ is the scalar _____. Since $\vec{A} \cdot \vec{A} = |\vec{A}|^2$, the magnitude of the vector $proj_{\vec{A}} \vec{B}$ is given by $|proj_{\vec{A}} \vec{B}| = \dfrac{|\vec{B} \cdot \vec{A}|}{} |\vec{A}| = \underline{\hspace{2cm}}.$

Thus, the absolute value of the scalar component of $\vec{B}$ in the direction $\vec{A}$ is simply the magnitude of _____.

22.   Consider the vectors $\vec{A} = -2\vec{i} + \vec{j} + 2\vec{k}$ and $\vec{B} = 2\vec{i} - 6\vec{j} - 3\vec{k}$. Then,
$$proj_{\vec{A}} \vec{B} = \left(\frac{\vec{B} \cdot \vec{A}}{\vec{A} \cdot \vec{A}}\right)\vec{A} = \frac{}{4 + 1 + 4} \vec{A} = \underline{\hspace{1cm}} \vec{A} = \underline{\hspace{2cm}}.$$

Then the scalar component of $\vec{B}$ in the direction $\vec{A}$ is the scalar $\dfrac{\vec{B} \cdot \vec{A}}{|\vec{A}|} = \underline{\hspace{1.5cm}}.$ The fact that this component is negative means that the vector $proj_{\vec{A}} \vec{B}$ points in the _____ direction of $\vec{A}$.

23.   For the vectors in Problem 22,
$$proj_{\vec{B}} \vec{A} = \left(\frac{\vec{A} \cdot \vec{B}}{\vec{B} \cdot \vec{B}}\right) \underline{\hspace{1.5cm}} = \frac{-16}{} \vec{B} = \underline{\hspace{2cm}}.$$

The scalar component of $\vec{A}$ in the direction of $\vec{B}$ is
$$\frac{\vec{A} \cdot \vec{B}}{} = \underline{\hspace{1.5cm}}.$$

OBJECTIVE C: Use vector methods to calculate the distance between a given point $P(x,y)$ and a given line L in the xy-plane.

24.   Find the distance between the point $P(1,-7)$ and the line $L:\ 5y - 7x + 14 = 0$.

---

20. $\left(\dfrac{\vec{B} \cdot \vec{A}}{\vec{A} \cdot \vec{A}}\right) \vec{A}$

21. $\dfrac{\vec{B} \cdot \vec{A}}{|\vec{A}|}$, $|\vec{A}|^2$, $\dfrac{|\vec{B} \cdot \vec{A}|}{|\vec{A}|}$, $proj_{\vec{A}} \vec{B}$

22. $-4 - 6 - 6$, $-\dfrac{16}{9}$, $\dfrac{32}{9}\vec{i} - \dfrac{16}{9}\vec{j} - \dfrac{32}{9}\vec{k}$, $-\dfrac{16}{3}$, opposite

23. $\vec{B}$, 49, $-\dfrac{32}{49}\vec{i} + \dfrac{96}{49}\vec{j} + \dfrac{48}{49}\vec{k}$, $|\vec{B}|$, $-\dfrac{16}{7}$

Solution. If we write the equation of the line L in the form $7x - 5y = 14$, we find a normal vector to the line is

$\vec{N}$ = _____. The point B(0,_____) lies on the line. Thus, the vector $\vec{BP}$ = _____ is a vector from the line L to P. The magnitude of the projection of $\vec{BP}$ onto the normal $\vec{N}$ gives the distance between P and L:

$$d = |\text{proj}_{\vec{N}}\, \vec{BP}| = \frac{|\underline{\quad\quad} \cdot \vec{N}|}{|\vec{N}|} = \frac{|\underline{\quad\quad}|}{\sqrt{74}} = \underline{\quad\quad}.$$

Notice that since $|\text{proj}_{\vec{N}}\, \vec{BP}|$ is the <u>absolute value</u> of the scalar component of $\vec{BP}$ in the direction $\vec{N}$, we need not be concerned with whether $\vec{BP}$ points in the same or opposite direction to $\vec{N}$: either way the procedure correctly computes a positive distance between P and L.

[OBJECTIVE D]: Know the following five properties of the scalar product.

25. $\vec{B} \cdot \vec{A}$ = _____   26. $\vec{A} \cdot (\vec{B} + \vec{C})$ = _____

27. $\vec{A} \cdot \vec{A} = 0$ if and only if _____.

28. $\vec{A} \cdot (c\vec{B}) = (c\vec{A}) \cdot \vec{B}$ = _____.

29. $\vec{A} \cdot \vec{A}$ is _____ for every nonzero vector $\vec{A}$.

[OBJECTIVE E]: Find an equation for the line in the xy-plane that passes through a given point and is perpendicular to a specified vector.

30. The line through the point P(-3,1) perpendicular to $\vec{N} = -\vec{i} + 4\vec{j}$ is

_____ = $(-1)(-3) + (4)(1)$

or

y = _____ .

---

24. $7\vec{i} - 5\vec{j}$, $-\frac{14}{5}$, $\vec{i} - \frac{21}{5}\vec{j}$, $\vec{BP}$, $7 + 21$, $\frac{14\sqrt{74}}{37}$ 25. $\vec{A} \cdot \vec{B}$

26. $\vec{A} \cdot \vec{B} + \vec{A} \cdot \vec{C}$ 27. $\vec{A} = \vec{0}$

28. $c(\vec{A} \cdot \vec{B})$ 29. positive

30. $-x + 4y$, $\frac{1}{4}(x + 7)$

11.4   CROSS PRODUCTS

OBJECTIVE A : Define the vector product or cross product of two
vectors in space, and give at least five properties
of the cross product.

31. $\vec{A} \times \vec{B} =$ (_____)$\vec{n}$, where $\vec{n}$ is a _____ vector
_____ to the plane of $\vec{A}$ and $\vec{B}$, and pointing in
the direction a right-threaded screw advances when its head is
rotated from ____ to ____ through the angle θ from $\vec{A}$
to $\vec{B}$.

32. $\vec{B} \times \vec{A} =$ _____.   33. $\vec{A} \times (\vec{B} + \vec{C}) =$ _____.

34. $(\vec{B} + \vec{C}) \times \vec{A} =$ _____.

35. $(r\vec{A}) \times (s\vec{B}) =$ _____.

36.   If $\vec{A} \times \vec{B} = \vec{0}$, then $\vec{A} = \vec{0}$ or $\vec{B} = \vec{0}$, or $\vec{A}$ and $\vec{B}$ are
_____ vectors.

37. $\vec{i} \times \vec{j} =$ ____, $\vec{j} \times \vec{k} =$ ____,   and $\vec{k} \times \vec{i} =$ ____.
These three equations are easily remembered by writing the
vectors in cyclic order $\vec{i}, \vec{j}, \vec{k}, \vec{i}, \vec{j}$ and realizing that the
cross product of any two results is the next one in line.

OBJECTIVE B : Use the determinant formula to calculate the cross
product of any two vectors in space whose $\vec{i}, \vec{j}, \vec{k}$
components are given.

38.   Let $\vec{A} = \vec{i} + \vec{j} + \vec{k}$ and $\vec{B} = -5\vec{i} + 3\vec{j} - 2\vec{k}$. Then

$$\vec{A} \times \vec{B} = \begin{vmatrix} \vec{i} & \vec{j} & \vec{k} \\ \rule{1em}{0.4pt} & \rule{1em}{0.4pt} & \rule{1em}{0.4pt} \\ \rule{1em}{0.4pt} & \rule{1em}{0.4pt} & \rule{1em}{0.4pt} \end{vmatrix} = (-2 - 3)\vec{i} + (-5 + \underline{\quad})\vec{j} + (\underline{\qquad})\vec{k}$$

$$= \underline{\hspace{4cm}}.$$

---

31. $|\vec{A}|\,|\vec{B}|\sin\theta$, unit, perpendicular, $\vec{A}$, $\vec{B}$ 32. $-\vec{A} \times \vec{B}$

33. $(\vec{A} \times \vec{B}) + (\vec{A} \times \vec{C})$ 34. $(\vec{B} \times \vec{A}) + (\vec{C} \times \vec{A})$

35. $rs(\vec{A} \times \vec{B})$ 36. parallel

37. $\vec{k}, \vec{i}, \vec{j}$ 38. $\begin{vmatrix} \vec{i} & \vec{j} & \vec{k} \\ 1 & 1 & 1 \\ -5 & 3 & -2 \end{vmatrix}$, 2, 3 + 5, $-5\vec{i} - 3\vec{j} + 8\vec{k}$

39. For the vectors in Problem 38, the magnitude $|\vec{A} \times \vec{B}| = $ _____
     represents the area of the _____ with sides _____
     and $\vec{B}$.

OBJECTIVE C : Find a vector that is perpendicular to two given
             vectors in space.

40. Let $\vec{A} = \vec{i} + \vec{k}$ and $\vec{B} = \vec{i} + 2\vec{j} - \vec{k}$. Then a vector
     perpendicular to both $\vec{A}$ and $\vec{B}$ is $\vec{A} \times \vec{B} = $ _____.

41. The three points $A(1,0,2)$, $B(3,1,4)$, and $C(-1,5,1)$
     determine a plane in space. The vectors $\vec{AB} = $ _____
     and $\vec{AC} = $ _____ both lie in that plane. Thus, the
     cross product $\vec{AB} \times \vec{AC} = $ _____ gives a vector that
     is perpendicular to the plane containing A, B, and C. Other
     vectors perpendicular to that plane are $\vec{CB} \times \vec{CA}$, $\vec{BA} \times \vec{BC}$,
     $\vec{AB} \times \vec{BC}$, etc.

OBJECTIVE D : Find the area of any triangle with specified vertices
             in space, and find the distance between the origin and
             the plane determined by that triangle.

42. The area of the triangle determined by the points in Problem 41
     is:

     area $= \frac{1}{2}$ _____ $= \frac{1}{2}\sqrt{121 + \underline{\quad} + \underline{\quad}} = $ _____.

43. To calculate the distance between the origin and the plane
     determined by the triangle in Problem 41, simply calculate the
     magnitude of the projection of any vector $\vec{OP}$ from the origin
     O to a point P on the plane onto a direction vector
     perpendicular to the plane. For instance, the point $A(1,0,2)$
     lies on the plane and $\vec{OA} = $ _____. A vector orthogonal to
     the plane is $\vec{N} = \vec{AB} \times \vec{AC}$ which was calculated in Problem 41.
     Hence,

     $|\text{proj}_{\vec{N}}\ \vec{OA}| = \dfrac{|\vec{OA} \cdot \underline{\quad}|}{|\vec{N}|} = \dfrac{|\underline{\qquad}|}{\sqrt{269}} = $ _____

     gives the distance between the origin and the plane of the
     triangle.

---

39. $7\sqrt{2}$, parallelogram, $\vec{A}$ 40. $-2\vec{i} + 2\vec{j} + \vec{k}$

41. $2\vec{i} + 2\vec{j} + 2\vec{k}$, $-2\vec{i} + 5\vec{j} - \vec{k}$, $-11\vec{i} - 2\vec{j} + 12\vec{k}$ 42. $|\vec{AB} \times \vec{AC}|$, 4, 144, $\frac{1}{2}\sqrt{269}$

43. $\vec{i} + 2\vec{k}$, $\vec{N}$, $-11 + 24$, $\dfrac{13}{\sqrt{269}} \approx .79262$

OBJECTIVE E : Given any three vectors $\vec{A}$, $\vec{B}$, and $\vec{C}$ in space, calculate the triple scalar product $\vec{A} \cdot (\vec{B} \times \vec{C})$.

44. For the vectors $\vec{A} = 3\vec{i} + 5\vec{j} - 2\vec{k}$, $\vec{B} = -\vec{i} + 4\vec{j} - 4\vec{k}$, and $\vec{C} = 4\vec{j} - \vec{k}$, the triple scalar product $\vec{A} \cdot (\vec{B} \times \vec{C})$ is given by the determinant

$$\vec{A} \cdot (\vec{B} \times \vec{C}) = \begin{vmatrix} \underline{} & \underline{} & \underline{} \\ \underline{} & \underline{} & \underline{} \\ \underline{} & \underline{} & \underline{} \end{vmatrix} = \underline{}.$$

45. $\vec{A} \cdot (\vec{B} \times \vec{C}) = (\vec{A} \times \underline{}) \cdot \underline{}$ so that the dot and the cross may be $\underline{}$ in the triple scalar product.

OBJECTIVE F : Find the volume of the box determined by three given vectors $\vec{A}$, $\vec{B}$, and $\vec{C}$.

46. For the three vectors given in Problem 44 above, the volume of the parallelepiped determined by them is the absolute value of the triple scalar product $\vec{A} \cdot (\underline{})$:

$$|\vec{A} \cdot (\underline{})| = \underline{} .$$

11.5   LINES AND PLANES IN SPACE

OBJECTIVE A : Write parametric equations of a line in space given (a) two points on the line, or (b) a point on the line and a vector parallel to the line.

47. Consider the line L determined by the two points $P(\frac{1}{2},0,1)$ and $Q(\frac{1}{2},-1,2)$. The vector $\vec{PQ} = \underline{}$ lies in a direction parallel to the line. Then, using the point P on the line,

$$x = \tfrac{1}{2} + 0t, \quad y = \underline{}, \quad \text{and} \quad z = \underline{}$$

are parametric equations for the line. Using the point Q,

$$x = \tfrac{1}{2} + 0s, \quad y = \underline{}, \quad \text{and} \quad z = \underline{}$$

is another set of parametric equations representing the same line. The point $(\frac{1}{2},-2,3)$ lies on the line when t = 2 and when s = $\underline{}$. The two sets of equations describe the same set of points (a straight line), but any particular point on that line will be given by different values of the parameters t and s in each representation.

---

44. $\begin{vmatrix} 3 & 5 & -2 \\ -1 & 4 & -4 \\ 0 & 4 & -1 \end{vmatrix}$, 39   45. $\vec{B}$, $\vec{C}$, interchanged

46. $\vec{B} \times \vec{C}$, $\vec{B} \times \vec{C}$, 39   47. $-\vec{j} + \vec{k}$, $0 - t$, $1 + t$, $-1 - s$, $2 + s$, 1

48. A line through the point (4,-2,2) in the direction parallel
     to the vector $\vec{A} = -2\vec{i} + \vec{j} + \vec{k}$ is given by x = _____,
     y = _____, and z = _____.

OBJECTIVE B: Find the distance between a given point and line in
               space.

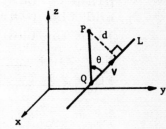

49. Suppose it is desired to find the
     distance between the point P(7,3,-1)
     and the line L: x = 4 - 2t,
     y = -2 + t, and z = 2 + t. The vector
     $\vec{v} = -2\vec{i} + \vec{j} + \vec{k}$ is _____ to L
     (see figure at the right). Choose any
     point on the line L, say Q(4,-2,2).
     Let θ denote the angle between $\vec{v}$ and
     $\vec{QP}$. Then, from the diagram at the right, $d = |\vec{QP}| \sin \theta$ and

     $|\vec{v}|d = |\vec{v}||\vec{QP}| \sin \theta =$ _____. Thus, $d = \dfrac{|\vec{v} \times \vec{QP}|}{|\vec{v}|}$ gives

     the desired distance. Now, $\vec{QP} =$ _____ and

     $$\vec{v} \times \vec{QP} = \begin{vmatrix} \vec{i} & \vec{j} & \vec{k} \\ -2 & 1 & 1 \\ \underline{\quad} & \underline{\quad} & \underline{\quad} \end{vmatrix} = \underline{\hspace{3cm}}.$$

     Hence, $d = \dfrac{\underline{\hspace{2cm}}}{\sqrt{4 + 1 + 1}} = \underline{\hspace{3cm}}.$

     Note: Problem 49 presents a different method from the text for
            finding the distance between a point and line in space.

OBJECTIVE C: Write an equation of a plane in space given (a) a
               point on the plane and a vector normal to the plane,
               or (b) three noncollinear points on the plane.

50. An equation of the plane in space through the point (1,-1,7)
     and perpendicular to the vector $\vec{v} = -2\vec{i} - 4\vec{j} + 5\vec{k}$ is given by
     -2(x - 1) + (-4)(_____) + _____(z - 7) = 0, or simplifying
     algebraically, _____ = -37.

51. To find an equation of the plane containing the three points
     A(4,-2,3), B(3,2,-1), and C(-1,1,1) observe that the vectors
     $\vec{AB} =$ _____ and $\vec{AC} =$ _____ both lie in the
     plane. Hence $\vec{AB} \times \vec{AC} =$ _____ is normal to the
     plane. Using the point A on the plane, an equation for it
     is 4(x - 4) + _____ (y + 2) + 17(_____) = 0, or
     _____ = 31.

---

48. 4 - 2t, -2 + t, 2 + t

49. parallel, $|\vec{v} \times \vec{QP}|$, $3\vec{i} + 5\vec{j} - 3\vec{k}$, 3, 5, -3, $-8\vec{i} - 3\vec{j} - 13\vec{k}$, $\sqrt{64 + 9 + 169}$, $\sqrt{\dfrac{121}{3}} = \dfrac{11}{3}\sqrt{3} \approx 6.35$

50. y + 1, 5, 2x + 4y - 5z   51. $-\vec{i} + 4\vec{j} - 4\vec{k}$, $-5\vec{i} + 3\vec{j} - 2\vec{k}$, $4\vec{i} + 18\vec{j} + 17\vec{k}$, 18, z - 3, 4x + 18y + 17z

OBJECTIVE D: Find the distance between a given point and plane in space.

52.   To find the distance between the point P and a plane, pick any point Q on the plane and calculate the magnitude of the projection vector of $\vec{PQ}$ onto a vector perpendicular to the plane. Let us find the distance between the point $P(5,-1,0)$ and the plane $2x + 4y - 5z = -37$. The point $Q(0,0,\underline{\quad})$ lies on the plane. A vector normal to the plane is

$\vec{N} = 2\vec{i} + \underline{\quad\quad}$, and we have the vector

$\vec{PQ} = \underline{\quad\quad\quad}$. Thus,

$$|\text{proj}_{\vec{N}}\ \vec{PQ}| = \frac{|\vec{PQ} \cdot \vec{N}|}{\underline{\quad\quad}} = \frac{|\underline{\quad}|}{\sqrt{45}} \approx 6.41$$

gives the distance between the point and the plane.

OBJECTIVE E: Find the point in which a given line meets a given plane.

53.   The line $x = 4 - 2t$, $y = -2 + t$, $z = 2 + t$ from Problem 48 meets the plane $x - 3y + 7z = 30$ when
$\underline{\quad\quad\quad\quad\quad\quad} = 30$ or $t = \underline{\quad\quad}$. The point of intersection is $P(\underline{\quad\quad})$.

OBJECTIVE F: Find parametric equations for the line in which two given nonparallel planes intersect. Find also the (acute) angle between the planes.

54.   A vector parallel to the line of intersection of the planes $-x + 2y - z = 4$ and $3x - y - 2z = 3$ is given by the cross product of their normals:

$$\begin{vmatrix} \vec{i} & \vec{j} & \vec{k} \\ -1 & 2 & -1 \\ 3 & -1 & -2 \end{vmatrix} = \underline{\quad\quad\quad\quad\quad} .$$

Substituting $x = 0$ in the plane equations and solving for y and z simultaneously gives $y = 1$ and $z = \underline{\quad\quad}$. Thus the point $(\underline{\quad\quad})$ is common to both planes and lies on the line of their intersection. The line is $x = \underline{\quad\quad}$, $y = 1 - 5t$, $z = \underline{\quad\quad}$.

---

52. $\frac{37}{5}$, $4\vec{j} - 5\vec{k}$, $-5\vec{i} + \vec{j} + \frac{37}{5}\vec{k}$, $|\vec{N}|$, -43

53. $(4 - 2t) - 3(-2 + t) + 7(2 + t)$, 3, $P(-2,1,5)$

54. $-5\vec{i} - 5\vec{j} - 5\vec{k}$, -2, $(0,1,-2)$, $-5t$, $-2 - 5t$

55. The (acute) angle between the two planes in Problem 54 is

$$\theta = \cos^{-1}\left(\frac{\vec{N}_1 \cdot \vec{N}_2}{|\vec{N}_1||\vec{N}_2|}\right)$$

$$= \cos^{-1}\left(\frac{-3 - 2 + 2}{\underline{\qquad}}\right)$$

$$= \cos^{-1}\left(\frac{-3}{2\sqrt{21}}\right)$$

$$= 71°. \text{(Calculator, rounded)}$$

11.6 SURFACES IN SPACE

OBJECTIVE A : Discuss and sketch cylinders whose **equations are given.**

56. Consider the surface given by
     y = sin x. It is a cylinder
     with elements parallel to the
     ____-axis. It extends indefinitely
     in both the positive and negative
     directions along the ____-axis.
     Sketch the graph in the coordinate
     system at the right.

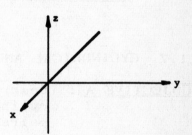

57. Consider the surface $z = x^3$.
     It is a cylinder with elements
     parallel to the ____-axis. It
     extends indefinitely in both the
     positive and negative directions
     along the y-axis, with parallel
     lines passing through the curve
     $z = x^3$ in the ____-plane.
     Sketch the graph at the right.

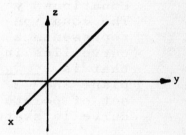

---

55. $\sqrt{6}\ \sqrt{14}$

56. z, z   57. y, xz

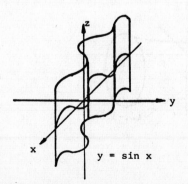

y = sin x

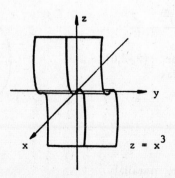

z = x³

OBJECTIVE B : Discuss and sketch a given surface whose equation
F(x,y,z) = 0 is a quadratic in the variables x, y,
and z.

58. Consider the surface given by the equation
$x^2 - 2y^2 + z^2 - 4x - 12y = 20$.
Completing the squares we obtain
$(x - 2)^2 +$ _____ = 6
or, dividing by 6,

_____.

This is a _____ of _____
sheet. It is centered at the point
_____. Cross sections
parallel to the plane y = 0 are
_____ while those parallel to
the other coordinate planes are
_____. Sketch the surface.

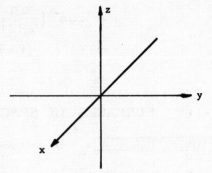

11.7 CYLINDRICAL AND SPHERICAL COORDINATES

OBJECTIVE A : Describe the set of points in space whose Cartesian,
cylindrical, or spherical coordinates satisfy given
pairs of simultaneous equations.

59. Consider the pair of Cartesian
equations $y^2 - z^2 = 1$, x = 2.
The equation $y^2 - z^2 = 1$
represents a _____. This
curve lies in the _____ x = 2
that is _____ to the yz-
plane and 2 units from it. The
set of points representing the
curve is sketched at the right.

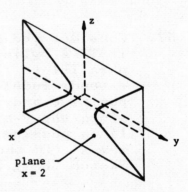

plane
x = 2

---

58. $-2(y + 3)^2 + z^2$,

$$\frac{(x - 2)^2}{6} - \frac{(y + 3)^2}{3} + \frac{z^2}{6} = 1,$$

hyperboloid, one, (2,-3,0),
circles, hyperbolas

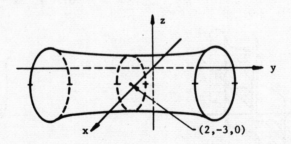

59. hyperbola, plane, parallel

60.   Consider the pair of cylindrical
equations z = 5,   r = θ. The
equation z = 5 represents a
_____ that is parallel to the
_____ and 5 units above it.
The equation r = θ represents a
_____ on that plane. It is
sketched at the right.

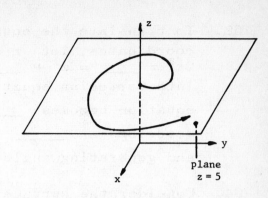

plane
z = 5

61.   Consider the pair of equations $\theta = \frac{\pi}{4}$, $\rho = 3 \sec \phi$
representing a set of points in spherical coordinates. The
equation $\rho = 3 \sec \phi$ can be written 3 = _____, or
3 = ____. Thus it represents a _____ parallel to the
_____ and 3 units above

it. The equation $\theta = \frac{\pi}{4}$
represents a _____ so
the set of points described by
the pair of equations is a half-

line or ray at an angle $\frac{\pi}{4}$
with the _____. It is sketched
at the right.

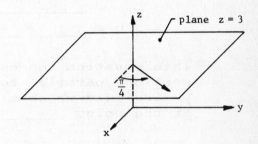

plane z = 3

OBJECTIVE B :   Translate an equation from a given coordinate system
(Cartesian, cylindrical, or spherical) into forms
that are appropriate to the other two systems.

62.   To translate the equation $z = x^2 + y^2$ into cylindrical
coordinates, let x = _____ and y = _____. Then,
z = ____ is an appropriate equation in the cylindrical
coordinate system. To translate this into spherical
coordinates, let z = _____ and r = _____ yielding
the equation _____ or $\cos \phi =$ _____.

---

60. plane, xy-plane, spiral

61. $\rho \cos \phi$, z, plane, xy-plane, half-plane, xz-plane, z = 3

62. $r \cos \theta$, $r \sin \theta$, r^2, $\rho \cos \phi$, $\rho \sin \phi$, $\rho \cos \phi = \rho \sin^2 \phi$, $\rho \sin^2 \phi$

63. To translate the equation $\phi = \frac{\pi}{4}$ into cylindrical
coordinates, let $r =$ _____ $=$ _____. Then
$2r^2 =$ ____ $= z^2 +$ ____, or $r^2 =$ ____. To translate this
into Cartesian equations, note that $r^2 =$ _____ so the
equation becomes _____. The equation $\phi = \frac{\pi}{4}$
describes a _____ with vertex at the origin 0, axis Oz,
and generating angle $\frac{\pi}{4}$.

64. Consider the surface given in
cylindrical coordinates by the
equation $6z = 12 + r^2 \cos 2\theta$.
Rewriting the equation in Cartesian
coordinates, we find

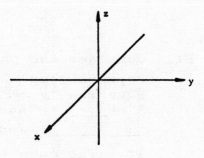

$6(z - 2) = r^2(\cos^2\theta -$ _____$)$

$\qquad = $ _____ $- r^2 \sin^2\theta$

$\qquad = $ _____ or,

$z - 2 = $ _____.

This equation represents a _____. Cross
sections parallel to the xy-plane are _____. If x or
y is fixed we obtain _____. The surface is centered
at the point _____. Sketch the surface.

---

63. $\rho \sin\phi$, $\frac{\sqrt{2}}{2}\rho$, ρ^2, r^2, z^2, $x^2 + y^2$, $z^2 = x^2 + y^2$, cone

64. $\sin^2\theta$, $r^2\cos^2\theta$,

$x^2 - y^2$, $\frac{x^2}{6} - \frac{y^2}{6}$,

hyperbolic paraboloid, hyperbolas,
parabolas, (0,0,2)

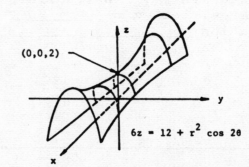

CHAPTER 11   SELF-TEST

1. Find unit vectors tangent and normal to the curve
$6y = x^3 - 6x^2 + 9x + 6$ at the point $P(0,1)$.

2. Given the vectors $\vec{A} = \vec{i} - \vec{j} + \vec{k}$ and $\vec{B} = -2\vec{j} + 3\vec{k}$:

   (a) Determine the cosine of the angle between $\vec{A}$ and $\vec{B}$.

   (b) Find $\vec{A} \times \vec{B}$.

   (c) Find the projection vector $\text{proj}_{\vec{A}}\, \vec{B}$,   and the component of
   $\vec{B}$ in the direction $\vec{A}$.

3. Find the distance between the point $P(5,0)$ in the xy-plane and
the line   L: $2y = -3x + 7$.

4. Find an equation for the line in the xy-plane that passes through
$(3,-5)$ and is perpendicular to the vector $\vec{N} = 2\vec{i} - 3\vec{j}$.

5. (a) Find an equation of the plane in space containing the points
   $A(\frac{1}{2},0,1)$, $B(-\frac{1}{2},1,-1)$,   and $C(-\frac{3}{2},2,1)$.

   (b) Find the area of the triangle with vertices   A, B, C.

   (c) Find the distance between the origin and the plane
   determined in part (a).

6. (a) Write parametric equations for the line in space through the
   points $(7,1,-2)$ and $(5,-1,0)$.

   (b) Write parametric equations for the line through the point
   $(12,0,-2)$ and parallel to the vector $\vec{v} = \vec{i} + \vec{j} - \vec{k}$.

7. Find the distance between the point $P(-3,1,0)$ and the line   L:
$x = 2 + t$, $y = -3 + 2t$, $z = 4 + 2t$.

8. Find the volume of the parallelepiped ("box") determined by
$\vec{u} = 2\vec{i} - \vec{j} + \vec{k}$, $\vec{v} = 3\vec{i} + 2\vec{j} - 2\vec{k}$,   and $\vec{w} = 3\vec{i} + 2\vec{j}$.

9. Find parametric equations for the line of intersection of the
planes $2x + z = 6$ and $y - 2z = -1$.

10. Sketch the cylinder $y = -\sin z$ in space.

11. Sketch the surface $z = \frac{1}{2}\sqrt{x^2 + y^2}$.

12. Sketch the quadric surface $z = x^2 + y^2 - 2x - 2y + 2$,   and
identify its type.

13. Describe the set of points whose cylindrical or spherical
coordinates satisfy the given pairs of simultaneous equations.
Sketch.
    (a) $\theta = \frac{\pi}{4}$, $z = r^2$ (b) $\phi = \frac{\pi}{3}$, $\rho = 2$

14. Translate the equation $r^2 = z(4 - z)$ from cylindrical into
spherical and Cartesian coordinates.

SOLUTIONS TO CHAPTER 11 SELF-TEST

1. Observe that the point P does lie on the curve since the coordinates of P satisfy the equation for the curve. Now, $y' = \frac{1}{2}(x^2 - 4x + 3)$ so that $y'(0) = \frac{3}{2}$. One vector having this slope is $\vec{v} = 2\vec{i} + 3\vec{j}$, so a tangent vector having this slope is $\frac{\vec{v}}{|\vec{v}|} = \frac{2}{\sqrt{13}}\vec{i} + \frac{3}{\sqrt{13}}\vec{j}$. A unit normal vector is $\vec{n} = -\frac{3}{\sqrt{13}}\vec{i} + \frac{2}{\sqrt{13}}\vec{j}$.

2. (a) $\cos\theta = \dfrac{\vec{A} \cdot \vec{B}}{|\vec{A}||\vec{B}|} = \dfrac{5}{\sqrt{3}\sqrt{13}} = \dfrac{5}{\sqrt{39}} \approx 0.80064$.

    (b) $\vec{A} \times \vec{B} = -\vec{i} - 3\vec{j} - 2\vec{k}$

    (c) $\text{proj}_{\vec{A}}\vec{B} = \left(\dfrac{\vec{B}\cdot\vec{A}}{\vec{A}\cdot\vec{A}}\right)\vec{A} = \frac{5}{3}\vec{i} - \frac{5}{3}\vec{j} + \frac{5}{3}\vec{k}$, and the component of $\vec{B}$ in the direction $\vec{A}$ is $\dfrac{\vec{B}\cdot\vec{A}}{|\vec{A}|} = \dfrac{5}{\sqrt{3}}$.

3. A normal vector to $3x + 2y = 7$ is $\vec{N} = 3\vec{i} + 2\vec{j}$. The point $B(0,\frac{7}{2})$ lies on the line. Then $\vec{BP} = 5\vec{i} - \frac{7}{2}\vec{j}$, and the distance is given by

    $$d = |\text{proj}_{\vec{N}}\vec{BP}| = \frac{|15 - 7|}{\sqrt{9 + 4}} = \frac{8}{\sqrt{13}} \approx 2.22 .$$

4. $2x - 3y = 2(3) + (-3)(-5) = 21$, or $y = \frac{2}{3}x - 7$.

5. (a) The vector $\vec{N} = \vec{AC} \times \vec{AB} = \begin{vmatrix} \vec{i} & \vec{j} & \vec{k} \\ -2 & 2 & 0 \\ -1 & 1 & -2 \end{vmatrix} = -4\vec{i} - 4\vec{j}$ is normal to the plane. An equation of the plane is given by

    $-4\left(x - \frac{1}{2}\right) - 4(y - 0) + 0(z - 1) = 0$ or, $4x + 4y = 2$.

    (b) area $= \frac{1}{2}|\vec{N}| = \frac{1}{2}\sqrt{16 + 16} = 2\sqrt{2}$.

    (c) The distance between the origin and the plane is given by

    $$d = |\text{proj}_{\vec{N}}\vec{OA}| = \frac{|\vec{OA}\cdot\vec{N}|}{|\vec{N}|} = \frac{|-2 + 0 + 0|}{4\sqrt{2}} = \frac{1}{2\sqrt{2}} \approx 0.354 .$$

6. (a) $x = 7 - 2t$, $y = 1 - 2t$, $z = -2 + 2t$

    (b) $x = 12 + t$, $y = t$, $z = -2 - t$

7. The vector $\vec{v} = \vec{i} + 2\vec{j} + 2\vec{k}$ is parallel to L, and the point Q(2,-3,4) lies on L. The vector from Q to P is given by $\vec{QP} = -5\vec{i} + 4\vec{j} - 4\vec{k}$. The distance between P and the line L is then given by

$$d = \frac{|\vec{v} \times \vec{QP}|}{|\vec{v}|} = \frac{|-16\vec{i} - 6\vec{j} + 14\vec{k}|}{|\vec{i} + 2\vec{j} + 2\vec{k}|} = \frac{\sqrt{488}}{\sqrt{9}} \approx 7.36.$$

8. The volume is the absolute value of the triple scalar product $\vec{u} \cdot (\vec{v} \times \vec{w})$. Thus,

$$\vec{u} \cdot (\vec{v} \times \vec{w}) = \begin{vmatrix} 2 & -1 & 1 \\ 3 & 2 & -2 \\ 3 & 2 & 0 \end{vmatrix} = 8 + (-1)(-6) + 1(6 - 6) = 14 ,$$

so the volume is 14 cubic units.

9. The vector

$$\vec{N_1} \times \vec{N_2} = \begin{vmatrix} \vec{i} & \vec{j} & \vec{k} \\ 2 & 0 & 1 \\ 0 & 1 & -2 \end{vmatrix} = -\vec{i} + 4\vec{j} + 2\vec{k}$$

is parallel to the line of intersection. Substituting z = 0 in the plane equations gives x = 3 and y = -1, so the point (3,-1,0) lies on the line. The line is given by

x = 3 - t,   y = -1 + 4t,   z = 2t .

10.

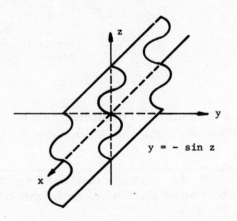

y = - sin z

11.

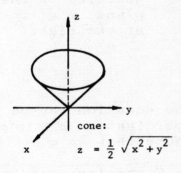

cone:

$z = \frac{1}{2} \sqrt{x^2 + y^2}$

12. Completing the squares,
$z = (x - 1)^2 + (y - 1)^2$,
which represents the circular
paraboloid sketched at the right.

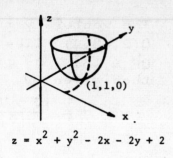

$z = x^2 + y^2 - 2x - 2y + 2$

13. (a) The equation $z = r^2$ represents
a parabola with vertical axis z
and horizontal axis r. The
parabola lies in the plane $\theta = \frac{\pi}{4}$
and is sketched at the right.

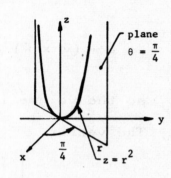

(b) The equation $\phi = \frac{\pi}{3}$ describes a
cone in spherical coordinates.
Also $z = \rho \cos \phi = 2 \cdot \frac{1}{2} = 1$
is constant. The intersection of
the plane $z = 1$ and the cone
$\phi = \frac{\pi}{3}$ results in a circle of
radius $r = \rho \sin \phi = \sqrt{3}$
centered on the z-axis in the
plane $z = 1$. It is sketched
at the right.

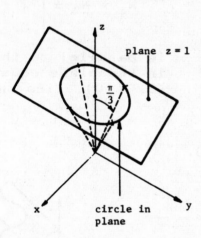

14. The equation can be written as $r^2 + z^2 = 4z$. In spherical
coordinates $\rho^2 \sin^2\phi + \rho^2 \cos^2\phi = 4\rho \cos \phi$ or, $\rho = 4 \cos \phi$.
(Note that $\rho = 0$ is included.)

In Cartesian coordinates, $x^2 + y^2 + z^2 = 4z$ or, completing the
square, $x^2 + y^2 + (z - 2)^2 = 4$, which represents a sphere of
radius 2 centered at the point $(0,0,2)$.

CHAPTER 12 VECTOR-VALUED FUNCTIONS AND MOTION IN SPACE

12.1 VECTOR-VALUED FUNCTIONS AND CURVES IN SPACE. DERIVATIVES AND INTEGRALS

1. A function $\vec{f} = f(t)\vec{i} + g(t)\vec{j} + h(t)\vec{k}$ is called a _____ function of t.

2. The most important vector function is the radius vector
$$\vec{r}(t) = f(t)\vec{i} + g(t)\vec{j} + h(t)\vec{k}$$

from the _____ to the point _____. This vector is also called the _____ vector of a particle moving through space.

OBJECTIVE A: Find the derivative of a given vector function and give the domain of the derived function.

3. If $\vec{f}(t) = f(t)\vec{i} + g(t)\vec{j} + h(t)\vec{k}$ is a vector function and each component is differentiable at $t = c$, then $\vec{f}$ is differentiable at $t = c$ and $\vec{f}' = $ _____.

4. Consider the vector function
$$\vec{f}(t) = \sqrt{1 - t^2}\,\vec{i} + \sinh t\,\vec{j} + 2^t\vec{k}.$$
The derivative is given by
$$\vec{f}'(t) = \underline{\hspace{5cm}}.$$

The first component, $\sqrt{1 - t^2}$, is differentiable for _____.
The second component, $\sinh t$, is differentiable for _____;
and the third component, 2^t, is differentiable for _____.
Thus, the vector function $\vec{f}$ is differentiable for

_____.

OBJECTIVE B: Given the position vector of a particle at any time t, find the velocity and acceleration vectors. Evaluate these vectors and find the speed and direction of motion of the particle at any instant of time.

---

1. vector   2. origin, $P\big(f(t), g(t), h(t)\big)$, position   3. $f'(c)\vec{i} + g'(c)\vec{j} + h'(c)\vec{k}$

4. $\dfrac{-t}{\sqrt{1 - t^2}}\vec{i} + \cosh t\,\vec{j} + 2^t \ln 2\,\vec{k}$, $-1 < t < 1$, all t, all t, $-1 < t < 1$

5.   The derivative $\dfrac{d\vec{r}}{dt}$ of the position vector is called the
_____ vector. It is designated by $\vec{v}$. The magnitude $|\vec{v}|$
is the _____ of the particle moving along the path
defined by _____. The direction of the particle's motion at
time t is the vector _____. The acceleration is the
vector _____.

6.   Consider the curve described by the position vector
$\vec{r}(t) = a\cos t\,\vec{i} + a\sin t\,\vec{j} + bt\,\vec{k}$, where a and b are
constants. The curve describes a circular helix. The velocity
and acceleration vectors are given by

$\vec{v} =$ _____ and $\vec{a} =$ _____,

respectively. When $t = \dfrac{\pi}{2}$, the velocity is $\vec{v}\left(\dfrac{\pi}{2}\right) =$ _____,
and the acceleration is $\vec{a}\left(\dfrac{\pi}{2}\right) =$ _____. The cosine of the angle
θ between the velocity and acceleration vectors when $t = \dfrac{\pi}{2}$ is

given by $\cos\theta = \dfrac{\vec{a}\cdot\vec{v}}{|\vec{a}||\vec{v}|} =$ _____. In fact, $\vec{a}\cdot\vec{v} =$ _____ for

all t, so that the velocity and acceleration vectors are always
perpendicular or orthogonal for the circular helix. Note that
the acceleration vector $\vec{a}$ always points in the direction from
the point $P(x,y,z)$ towards the z-axis in this example,
because $a\cos t\,\vec{i} + a\sin t\,\vec{j}$ describes a circular path of
radius a about the z-axis.

7.   The speed of the particle moving along the curve in Problem 6
at any time t is $|\vec{v}| =$ _____. The
direction of motion is the vector _____.

8.   If $\vec{r}(t)$ gives the position of a particle moving along a path,
and t is a differentiable function of the variable s, then
according to the chain rule

$$\frac{d\vec{r}}{ds} = \underline{\hspace{3cm}}.$$

9.   If $\vec{r}(t) = e^t\,\vec{i} + \sin t\,\vec{j} + t^2\,\vec{k}$, and $t = \sqrt{s}$ for $s > 0$, then

$$\frac{d\vec{r}}{ds} = (e^t\,\vec{i} + \underline{\hspace{1.5cm}}\,\vec{j} + 2t\,\vec{k})\cdot\underline{\hspace{1.5cm}}$$

$$= \frac{e^{\sqrt{s}}}{2\sqrt{s}}\,\vec{i} + \underline{\hspace{1.5cm}}\,\vec{j} + \vec{k}.$$

---

5.   velocity, speed, $\vec{r}$, $|\vec{v}|\,\vec{v}$, $\dfrac{d\vec{v}}{dt}$ or $\dfrac{d^2\vec{r}}{dt^2}$

6. $-a\sin t\,\vec{i} + a\cos t\,\vec{j} + b\,\vec{k}$, $-a\cos t\,\vec{i} - a\sin t\,\vec{j}$, $-a\vec{i} + b\vec{k}$, $-a\vec{j}$,   zero,   0

7. $\sqrt{a^2 + b^2}$, $\sqrt{a^2 + b^2}(-a\sin t\,\vec{i} + a\cos t\,\vec{j} + b\vec{k})$

8. $\dfrac{d\vec{r}}{dt}\dfrac{dt}{ds}$

9. $\cos t$, $\dfrac{1}{2\sqrt{s}}$, $\dfrac{\cos\sqrt{s}}{2\sqrt{s}}$

: Using the appropriate rules, calculate the derivatives of vector expressions involving the dot or cross products.

10. $\frac{d}{dt}(\vec{u} \cdot \vec{v}) = $ _____ .

11. $\frac{d}{dt}(\vec{u} \times \vec{v}) = $ _____ .

12. Suppose that $\vec{u}$ is a twice differentiable vector function of t. Then

$$\frac{d}{dt}\left(\vec{u} \times \frac{d\vec{u}}{dt}\right) = \frac{d\vec{u}}{dt} \times \frac{d\vec{u}}{dt} + \text{_____} = \text{\_\_} \times \frac{d^2\vec{u}}{dt^2}.$$

Therefore, if $\vec{u}$ is parallel to $\frac{d^2\vec{u}}{dt^2}$ for all t, $\frac{d}{dt}\left(\vec{u} \times \frac{d\vec{u}}{dt}\right)$

is the _____ vector. It follows that $\vec{u} \times \frac{d\vec{u}}{dt}$ is a _____

vector since the derivative of each component is zero (so that each component is a constant scalar function).

13. If $\vec{u}$ is a differentiable vector function of constant length, then $\vec{u}$ and $\frac{d\vec{u}}{dt}$ are _____ to each other.

: Find the definite or indefinite integral of a vector-valued function having continuous components.

14. $\int \left[t^3\vec{i} + e^{-2t}\vec{j} - \frac{1}{3t}\vec{k} \right] dt$

$= $ _____ $+ \vec{c}$, $t \neq 0$.

Here $\vec{c}$ is an arbitrary constant _____ .

: Find the position vector $\vec{r}(t)$ when the acceleration vector $\vec{a}(t)$ is known together with initial conditions $\vec{v}(0)$ and $\vec{r}(0)$. That is, solve a vector-valued initial value problem.

15. Suppose $\vec{a} = e^t\vec{i} + e^{-t}\vec{j}$ with $\vec{v}(0) = -\vec{j}$ and $\vec{r}(0) = 2\vec{i} - \vec{j}$. Note that the $\vec{k}$-component is zero and the particles moves in the xy-plane. Integration of $\vec{a}$ gives the velocity vector

$$\vec{v} = \text{_____} + \vec{c}_1$$

where $\vec{c}_1$ is a constant vector. From the initial condition $\vec{v}(0) = -\vec{j}$ we find

$$-\vec{j} = \text{_____} + \vec{c}_1$$

---

10. $\frac{d\vec{u}}{dt} \cdot \vec{v} + \vec{u} \cdot \frac{d\vec{v}}{dt}$ 11. $\frac{d\vec{u}}{dt} \times \vec{v} + \vec{u} \times \frac{d\vec{v}}{dt}$ 12. $\vec{u} \times \frac{d^2\vec{u}}{dt^2}$, $\vec{u}$, zero, constant

13. orthogonal   14. $\frac{1}{4}t^4\vec{i} - \frac{1}{2}e^{-2t}\vec{j} - \frac{1}{3}\ln|t|\vec{k}$, vector

or, solving for the constant vector,

$$\vec{c}_1 = \underline{\hspace{3cm}} .$$

Thus,

$$\vec{v} = (e^t - 1)\vec{i} - e^{-t}\vec{j} .$$

A second integration gives

$$\vec{r} = \underline{\hspace{2cm}} \vec{i} + e^{-t}\vec{j} + \vec{c}_2$$

where $\vec{c}_2$ is a constant vector. From the initial condition $\vec{r}(0) = 2\vec{i} - \vec{j}$ we find

$$2\vec{i} - \vec{j} = \underline{\hspace{3cm}} + \vec{c}_2$$

or $\vec{c}_2 = \vec{i} - 2\vec{j}$. Thus,

$$\vec{r}(t) = (e^t - t + 1)\vec{i} + \underline{\hspace{2cm}}\vec{j} .$$

12.2   MODELING PROJECTILE MOTION

OBJECTIVE : Use the parametric equations describing the motion of a projectile fired with initial speed v_0 at an angle of elevation α to answer questions concerning the motion of the projectile. Assume that air resistance is neglected and gravity is the only force acting on the projectile.

16.   For a projectile fired with an initial speed v_0 ft/sec at an angle of elevation α, the parametric equations for its motion are $x = c_1 t + \underline{\hspace{1.5cm}}$ ft and $y = \underline{\hspace{2cm}} + c_3 t + c_4$ ft, where c_1, c_2, c_3, and c_4 are constants determined from the initial conditions. Here t is measured in $\underline{\hspace{2cm}}$.

17.   The initial conditions give the $\underline{\hspace{5cm}}$ $x(t_0)$ and $y(t_0)$, and the $\underline{\hspace{4cm}}$ $\frac{dx}{dt}(t_0)$ and $\frac{dy}{dt}(t_0)$ at some given instant $t = t_0$.

18.   Suppose a boy throws a ball at an angle of $45°$ with the horizontal and with an initial speed of 30 ft/sec toward a building 35 ft away. If the ball leaves the boy's hand from a height of 3 ft above the ground, will it reach the building?

Solution.   For this problem, the initial conditions are $t = 0$ sec, $x = \underline{\hspace{1.5cm}}$ ft, $y = \underline{\hspace{1.5cm}}$ ft, $\frac{dx}{dt}(0) = 30 \cos \underline{\hspace{1.5cm}}$ ≈ 21.21 ft/sec, and $\frac{dy}{dt}(0) = \underline{\hspace{1.5cm}} \sin 45° \approx \underline{\hspace{1.5cm}}$ ft/sec.

---

15. $e^t\vec{i} - e^{-t}\vec{j}$, $\vec{i} - \vec{j}$, $-\vec{i}$, $e^t - t$, $\vec{i} + \vec{j}$, $e^t - 2$

16. c_2, $-\frac{1}{2}gt^2$, seconds   17.   position coordinates, velocity components

The parametric equations for the motion are $x = c_1 t + c_2$ and $y = -\frac{1}{2}gt^2 + c_3 t + c_4$. From the initial conditions, we find $c_2 = 0$, $c_1 = $ _____ , $c_4 = $ _____ , and $c_3 = 21.21$. To find the horizontal range R, we set $y = -\frac{1}{2}gt^2 + 21.21t + 3$ equal to _____ and solve for t: using $g = 32$ ft/sec^2,

$$t \approx \frac{-21.21 \pm \sqrt{(21.21)^2 + \underline{\quad\quad}}}{-32} \approx 1.45 \text{ sec}$$

(since t must be nonnegative we take the negative square root). Then, $R = $ _____ $\cdot (1.45)$, the value of _____ when $t = 1.45$. Hence, $R \approx$ _____ ft. Since $R < 35$ ft, the ball _____ reach the building.

12.3   DIRECTED DISTANCE AND THE UNIT TANGENT VECTOR $\vec{T}$

[OBJECTIVE A]:   Given the coordinates for a curve in space in terms of some parameter t, find the length of the curve for a specified interval $a \leq t \leq b$.

19.   Suppose the curve is given by $x = \frac{1}{3}(1 + t)^{3/2}$, $y = \frac{1}{3}(1 - t)^{3/2}$ and $z = \frac{1}{2}t$. To find the length of the curve for $-1 \leq t \leq 1$, calculate

$$\frac{dx}{dt} = \underline{\quad\quad\quad}, \quad \frac{dy}{dt} = \underline{\quad\quad\quad}, \quad \frac{dz}{dt} = \frac{1}{2}.$$

Then, $\dfrac{ds}{dt} = \sqrt{\left(\dfrac{dx}{dt}\right)^2 + \left(\dfrac{dy}{dt}\right)^2 + \left(\dfrac{dz}{dt}\right)^2}$

$$= \sqrt{\underline{\quad\quad\quad\quad} + \frac{1}{4}} = \underline{\quad\quad\quad}.$$

The length of the given curve is therefore

$$s = \int_{-1}^{1} \left(\frac{ds}{dt}\right) dt = \frac{1}{2}\sqrt{3} \int_{-1}^{1} \underline{\quad\quad} dt = \underline{\quad\quad}.$$

[OBJECTIVE B]:   Given the position vector for the motion of a particle, find the unit tangent vector $\vec{T}$ to the curve at any point of the curve.

20.   Suppose the position vector of the motion in the xy-plane is given by $\vec{r} = \ln t\,\vec{i} + t^2\vec{j}$ for $t > 0$. Then

---

18.   0, 3, 45°, 30, 21.21, 21.21, 3, zero, 192, 21.21, x, 30.75, **does not**

19. $\frac{1}{2}(1 + t)^{1/2}$, $-\frac{1}{2}(1 - t)^{1/2}$, $\frac{1}{4}(1 + t) + \frac{1}{4}(1 - t)$, $\frac{1}{2}\sqrt{3}$, 1, $\sqrt{3}$

$$\vec{v} = \frac{d\vec{r}}{dt} = \underline{\hspace{3cm}} \; ; \quad |\vec{v}| = \sqrt{\underline{\hspace{3cm}}} = \frac{1}{t}\sqrt{\underline{\hspace{2cm}}} \; ;$$

$$\vec{T} = \frac{\vec{v}}{|\vec{v}|} = \left(1 + 4t^4\right)^{-1/2}\vec{i} + \underline{\hspace{4cm}}\vec{j}.$$

21. Consider the motion in space described by the position vector
$\vec{r} = \tan t\,\vec{i} + \vec{j} + \sec t\,\vec{k}$, for $-\frac{\pi}{2} < t < \frac{\pi}{2}$. Then,

$$\vec{v} = \frac{d\vec{r}}{dt} = \underline{\hspace{6cm}} \; ;$$

$$|\vec{v}| = \sqrt{\sec^4 t + \underline{\hspace{4cm}}}$$

$$= |\sec t|\,\sqrt{\underline{\hspace{4cm}}} = (\sec t)\sqrt{2\sec^2 t - 1} \; ;$$

$$\vec{T} = \frac{\vec{v}}{|\vec{v}|} = \underline{\hspace{8cm}} \; .$$

22. Suppose the tangent vector to a curve in the xy-plane, having
position vector $\vec{r}(t)$, is everywhere perpendicular to some
nonzero constant vector $\vec{a}$. Then,

$$\frac{d\vec{r}}{dt} = f(t)\vec{u} \; ,$$

where $\vec{u}$ is a unit vector perpendicular to $\vec{a}$. Let $g(t)$ be
an indefinite integral of f, so that $\frac{d}{dt}g(t) = \underline{\hspace{1.5cm}}$. Then,

$$\frac{d}{dt}[\vec{r}(t) - g(t)\vec{u}] = \underline{\hspace{3cm}} = \vec{0} \; .$$

It follows that $\vec{r}(t) - g(t)\vec{u}$ is a constant vector $\vec{c}$, so
that $\vec{r}(t) = \underline{\hspace{2cm}}$. Therefore, the curve must lie along
the line $\vec{c} + s\vec{u}$, where $s = g(t)$. Thus, if the tangent
vector to a curve is everywhere perpendicular to a constant
nonzero vector $\vec{a}$, then the curve runs along a straight line
perpendicular to $\vec{a}$.

12.4   CURVATURE, TORSION, AND THE $\vec{T}\vec{N}\vec{B}$ FRAME

OBJECTIVE A : Given a curve in the xy-plane, find the curvature at
any point on the curve.

23. If $\vec{r} = x\vec{i} + y\vec{j}$, where the scalar components x and y are
differentiable functions of t, then the curvature κ may be
computed according to the formula

---

20. $\frac{1}{t}\vec{i} + 2t\vec{j}$, $\frac{1}{t^2} + 4t^2$, $1 + 4t^4$, $2t^2\left(1 + 4t^4\right)^{-1/2}$

21. $\sec^2 t\,\vec{i} + \sec t \tan t\,\vec{k}$, $\sec^2 t \tan^2 t$, $\sec^2 t + \tan^2 t$, $\sec t\,(2\sec^2 t - 1)^{-1/2}\vec{i} + \tan t\,(2\sec^2 t - 1)^{-1/2}\vec{k}$

22. $f(t)$, $\frac{d\vec{r}}{dt} - f(t)\vec{u}$, $\vec{c} + g(t)\vec{u}$

$$\kappa = \frac{}{|\vec{v}|^3} = \frac{|\dot{x}\ddot{y} - \ddot{x}\dot{y}|}{\left(\dot{x}^2 + \dot{y}^2\right)^{3/2}} .$$

Note that κ is a scalar quantity, and we assume that the velocity vector $\vec{v}$ is not _____.

24. Suppose a curve is given parametrically by $x = t$ and $y = \frac{1}{3}t^3$. Then, $\dot{x} = $ ____, $\dot{y} = $ ____, $\ddot{x} = $ ____, $\ddot{y} = $ ____. Next calculate

$$\vec{v} \times \vec{a} = \underline{\hspace{3cm}}\vec{k} = 2t\vec{k} .$$

Hence,

$$\kappa = \frac{|2t|}{\underline{\hspace{1.5cm}}} .$$

25. Given the curve $y = \frac{1}{x}$, $x \neq 0$, first parameterize it in the form $x = t$ and $y = \frac{1}{t}$, $t \neq 0$. The position vector is $\vec{r} = \underline{\hspace{3cm}}$. Then $\vec{v} = \underline{\hspace{3cm}}$ and $\vec{a} = -\frac{2}{t^3}\vec{j}$. Thus,

$$\vec{v} \times \vec{a} = \begin{vmatrix} \vec{i} & \vec{j} & \vec{k} \\ 1 & -1/t^2 & 0 \\ 0 & -2/t^3 & 0 \end{vmatrix}$$

Hence the curvature is

$$\kappa = \frac{|2/t^3|}{\left[\underline{\hspace{2cm}}\right]^{3/2}} = \underline{\hspace{3cm}} .$$

OBJECTIVE B : Given a curve in the xy-plane, find an equation of the circle of curvature at a specified point.

26. Let us find an equation of the osculating circle to $y = \sin x$ at the point $\left(\frac{\pi}{2}, 1\right)$. First we parameterize the curve: $\vec{r} = t\vec{i} + \underline{\hspace{1.5cm}}\vec{j}$. Then the velocity and acceleration vectors are $\vec{v} = \vec{i} + \cos t\vec{j}$ and $\vec{a} = \underline{\hspace{2cm}}$. Thus,

---

23. $|\vec{v} \times \vec{a}|$, zero   24. 1, t^2, 0, $2t$, $|\dot{x}\ddot{y} - \ddot{x}\dot{y}|$, $\left(1 + t^4\right)^{3/2}$

25. $t\vec{i} + \frac{1}{t}\vec{j}$, $\vec{i} - \frac{1}{t^2}\vec{j}$, $-\frac{2}{t^3}\vec{k}$, $1 + \frac{1}{t^4}$, $\frac{|2x^3|}{\left(1 + x^4\right)^{3/2}}$ since $t = x$

$$\vec{v} \times \vec{a} = \begin{vmatrix} \vec{i} & \vec{j} & \vec{k} \\ 1 & \cos t & 0 \\ 0 & -\sin t & 0 \end{vmatrix} = \underline{\hspace{3cm}}.$$

The curvature is

$$\kappa = \frac{|\sin t|}{[\underline{\hspace{2cm}}]^{3/2}}$$

so that at $t = \frac{\pi}{2}$, $\kappa = \underline{\hspace{2cm}}$. The radius of curvature is $\rho = \frac{1}{\kappa} = \underline{\hspace{2cm}}$ which is the radius of the osculating circle. The tangent to $y = \sin x$ when $x = \frac{\pi}{2}$ is horizontal, so the normal is in the $\underline{\hspace{2cm}}$ direction. Therefore, the center of the osculating circle is the point $\underline{\hspace{2cm}}$. An equation of the circle is given by

$$\underline{\hspace{5cm}}.$$

OBJECTIVE C : Given the coordinates of a curve in space as differen-
             tiable functions of t, find the curvature at any
             point using the formula $\kappa = \dfrac{|\vec{v} \times \vec{a}|}{|\vec{v}|^3}$.

27. Consider the space curve with position described by
    $\vec{r} = (t + 1)\vec{i} - t^2\vec{j} + (1 - 2t)\vec{k}$. To calculate the curvature κ
    at any point, we find $\vec{v} = \underline{\hspace{4cm}}$ and $\vec{a} = \underline{\hspace{2cm}}$.
    Hence,

$$\vec{v} \times \vec{a} = \begin{vmatrix} \vec{i} & \vec{j} & \vec{k} \\ 1 & -2t & -2 \\ 0 & -2 & 0 \end{vmatrix} = \underline{\hspace{3cm}}, \quad \text{and}$$

$$|\vec{v} \times \vec{a}| = \sqrt{\underline{\hspace{3cm}}} = \underline{\hspace{2cm}}. \quad \text{Also,} \quad |\vec{v}| = \underline{\hspace{3cm}}.$$

Therefore, $\kappa = \dfrac{|\vec{v} \times \vec{a}|}{\underline{\hspace{1cm}}} = \underline{\hspace{4cm}}$ gives the curvature for

any value of t. In particular, for $t = 0$, $\kappa = \underline{\hspace{3cm}}$.

OBJECTIVE D : Given the coordinates of a curve in space as differen-
             tiable functions of t, find the torsion τ using
             Formula 16 on page 801 of Finney/Thomas.

---

26. $\sin t$, $-\sin t\vec{j}$, $-\sin t\vec{k}$, $(1 + \cos^2 t)$, 1, 1, vertical, $\left(\frac{\pi}{2}, 0\right)$, $\left(x - \frac{\pi}{2}\right)^2 + y^2 = 1$

27. $\vec{i} - 2t\vec{j} - 2\vec{k}$, $-2\vec{j}$, $-4\vec{i} - 2\vec{k}$, $16 + 4$, $2\sqrt{5}$, $\sqrt{5 + 4t^2}$, $|\vec{v}|^3$, $\dfrac{2\sqrt{5}}{\left(5 + 4t^2\right)^{3/2}}$, $\dfrac{2}{5}$

28. For the curve $\vec{r} = (t + 1)\vec{i} - t^2\vec{j} + (1 - 2t)\vec{k}$ in Problem 27, we calculate the determinant

$$\begin{vmatrix} \dot{x} & \dot{y} & \dot{z} \\ \ddot{x} & \ddot{y} & \ddot{z} \\ \dddot{x} & \dddot{y} & \dddot{z} \end{vmatrix} = \begin{vmatrix} \underline{\quad} & \underline{\quad} & \underline{\quad} \\ \underline{\quad} & \underline{\quad} & \underline{\quad} \\ 0 & 0 & 0 \end{vmatrix} = \underline{\quad}.$$

Therefore, the torsion $\tau = 0$ so the moving particle tends to remain in the osculating plane as it travels along the curve.

OBJECTIVE E : Given the coordinates for a curve in terms of some parameter t, find the unit tangent vector $\vec{T}$, the principal normal vector $\vec{N}$, the curvature κ, the unit binormal vector $\vec{B}$, and the torsion τ.

29. Suppose the position vector of the curve is given by

$\vec{r} = 2t\vec{i} + t^2\vec{j} + \frac{1}{3}t^3\vec{k}$. Then $\vec{v} = \dfrac{d\vec{r}}{dt} = 2\vec{i} + 2t\vec{j} + \underline{\quad}\vec{k}$ and

$|\vec{v}| = \sqrt{\underline{\hspace{3cm}}} = \sqrt{\left(2 + t^2\right)^2} = 2 + t^2$,

$\vec{T} = \dfrac{\vec{v}}{|\vec{v}|} = \dfrac{2}{2 + t^2}\vec{i} + \underline{\hspace{2.5cm}}\vec{j} + \dfrac{t^2}{2 + t^2}\vec{k}$.

Differentiation of $\vec{T}$ with respect to t gives

$\dfrac{d\vec{T}}{dt} = -\dfrac{4t}{\left(2 + t^2\right)^2}\vec{i} + \underline{\hspace{2.5cm}}\vec{j} + \dfrac{4t}{\left(2 + t^2\right)^2}\vec{k}$

$= \dfrac{2}{\left(2 + t^2\right)^2}[\underline{\hspace{4cm}}]$.

By the rule $\dfrac{ds}{dt} = |\vec{v}|$ and the chain rule,

$\dfrac{d\vec{T}}{ds} = \dfrac{d\vec{T}/dt}{ds/dt} = \underline{\hspace{2cm}} [-2t\vec{i} + (2 - t^2)\vec{j} + 2t\vec{k}]$;

$\kappa = \left|\dfrac{d\vec{T}}{ds}\right| = \dfrac{2}{\left(2 + t^2\right)^3}\sqrt{\underline{\hspace{4cm}}} = \dfrac{2}{\left(2 + t^2\right)^2}$.

Then, $\vec{N} = \dfrac{d\vec{T}/ds}{|d\vec{T}/ds|} = -\dfrac{2t}{2 + t^2}\vec{i} + \underline{\hspace{1.5cm}}\vec{j} + \underline{\hspace{1.5cm}}\vec{k}$.

Finally, $\vec{B} = \vec{T} \times \vec{N}$ gives

$\vec{B} = \dfrac{t^2}{2 + t^2}\vec{i} + \underline{\hspace{1.5cm}}\vec{j} + \dfrac{2}{2 + t^2}\vec{k}$.

28. $\begin{vmatrix} 1 & -2t & -2 \\ 0 & -2 & 0 \\ 0 & 0 & 0 \end{vmatrix}$, 0

29. t^2, $4 + 4t^2 + t^4$, $\dfrac{2t}{2 + t^2}$, $\dfrac{2(2 - t^2)}{\left(2 + t^2\right)^2}$, $-2t\vec{i} + (2 - t^2)\vec{j} + 2t\vec{k}$, $\dfrac{2}{\left(2 + t^2\right)^3}$,

$4t^2 + \left(2 - t^2\right)^2 + 4t^2$, $\dfrac{2 - t^2}{2 + t^2}$, $\dfrac{2t}{2 + t^2}$, $\dfrac{-2t}{2 + t^2}$

30.   For the curve in Problem 29,

$$\begin{vmatrix} \dot{x} & \dot{y} & \dot{z} \\ \ddot{x} & \ddot{y} & \ddot{z} \\ \dddot{x} & \dddot{y} & \dddot{z} \end{vmatrix} =$$

$$\begin{vmatrix} \underline{\quad} & \underline{\quad} & \underline{\quad} \\ \underline{\quad} & \underline{\quad} & \underline{\quad} \\ \underline{\quad} & \underline{\quad} & \underline{\quad} \end{vmatrix} = \underline{\hspace{2cm}} \; .$$

Also,

$$\vec{v} \times \vec{a} = \begin{vmatrix} \vec{i} & \vec{j} & \vec{k} \\ 2 & 2t & t^2 \\ 0 & 2 & 2t \end{vmatrix} = \underline{\hspace{4cm}} \; .$$

Therefore, the torsion is

$$\tau = \dfrac{8}{\underline{\hspace{2cm}}} \; .$$

[OBJECTIVE F]:   Given the position vector for the motion of a particle,
find the velocity and acceleration vectors, the speed
$\dfrac{ds}{dt}$, and the tangential and normal components of
acceleration.

31.   Consider the path described by $\vec{r} = t\vec{i} + \sin t\vec{j}$. Then,

$\vec{v} = \dfrac{d\vec{r}}{dt} = \underline{\hspace{2.5cm}}$, the velocity vector;

$\vec{a} = \dfrac{d\vec{v}}{dt} = \underline{\hspace{2.5cm}}$, the acceleration vector;

$\dfrac{ds}{dt} = |\vec{v}| = \underline{\hspace{2cm}}$, the speed;

$\dfrac{d^2s}{dt^2} = a_T = \dfrac{\underline{\hspace{2cm}}}{\sqrt{1 + \cos^2 t}}$, the tangential component of
acceleration;

---

30. $\begin{vmatrix} 2 & 2t & t^2 \\ 0 & 2 & 2t \\ 0 & 0 & 2 \end{vmatrix}$, 8, $2t^2\vec{i} - 4t\vec{j} + 4\vec{k}$, $4t^4 + 16t^2 + 16$

and

$$a_N = \sqrt{|\vec{a}|^2 - \underline{\hspace{1cm}}} = \sqrt{\sin^2 t - \underline{\hspace{2cm}}}$$

$$= \sqrt{\frac{\overline{\hspace{1cm}}}{1 + \cos^2 t}}, \quad \text{or} \quad a_N = \frac{|\sin t|}{\sqrt{1 + \cos^2 t}}, \quad \begin{array}{l}\text{the normal component} \\ \text{of acceleration.}\end{array}$$

32. For the path given by $\vec{r} = 3\vec{i} + t^2\vec{j} + t^3\vec{k}, \quad t \geq 0,$

$\vec{v} = \dfrac{d\vec{r}}{dt} = \underline{\hspace{2cm}}$, the velocity vector;

$\vec{a} = \dfrac{d\vec{v}}{dt} = \underline{\hspace{2cm}}$, the acceleration vector;

$\dfrac{ds}{dt} = |\vec{v}| = \underline{\hspace{2cm}}$, the speed;

$\dfrac{d^2s}{dt^2} = \sqrt{4 + 9t^2} + \dfrac{\overline{\hspace{1cm}}}{\sqrt{4 + 9t^2}} = \dfrac{\overline{\hspace{1cm}}}{\sqrt{4 + 9t^2}}$, the tangential component of acceleration; and

$$a_N = \sqrt{\underline{\hspace{1.5cm}} - a_T^2} = \sqrt{4 + 36t^2 - \frac{\left(4 + 18t^2\right)^2}{4 + 9t^2}}$$

$$= \frac{\sqrt{\left(4 + 36t^2\right)\underline{\hspace{1cm}} - \left(4 + 18t^2\right)^2}}{\sqrt{4 + 9t^2}} = \frac{\overline{\hspace{1cm}}}{\sqrt{4 + 9t^2}}, \quad \begin{array}{l}\text{the normal} \\ \text{component of} \\ \text{acceleration.}\end{array}$$

Remark: If $t < 0,$ then $|\vec{v}| = -t\sqrt{4 + 9t^2}.$

12.5 PLANETARY MOTION AND SATELLITES

OBJECTIVE :   Given a curve $r = f(\theta)$ in the plane in polar coordinates and the rate $\dfrac{d\theta}{dt}$, or given the polar coordinates r and θ as functions of the parameter t, find the velocity and acceleration vectors in terms of $\vec{u}_r$ and $\vec{u}_\theta.$

33. In terms of the vectors $\vec{u}_r$ and $\vec{u}_\theta$, $\vec{v} = \underline{\hspace{1cm}} \vec{u}_r + \underline{\hspace{1cm}} \vec{u}_\theta$ and $\vec{a} = [\underline{\hspace{3cm}}]\vec{u}_r + [\underline{\hspace{3cm}}]\vec{u}_\theta$ give the velocity and acceleration vectors, respectively, of the moving particle in polar coordinates.

---

31. $\vec{i} + \cos t\vec{j}$, $-\sin t\vec{j}$, $\sqrt{1 + \cos^2 t}$, $-\sin t \cos t$, a_T^2, $\dfrac{\sin^2 t + \cos^2 t}{1 + \cos^2 t}$, $\sin^2 t$

32. $2t\vec{j} + 3t^2\vec{k}$, $2\vec{j} + 6t\vec{k}$, $t\sqrt{4 + 9t^2}$, $9t^2$, $4 + 18t^2$, $|\vec{a}|^2$, $4 + 9t^2$, $6t$

33. $\dfrac{dr}{dt}$, $r\dfrac{d\theta}{dt}$, $\dfrac{d^2r}{dt^2} - r\left(\dfrac{d\theta}{dt}\right)^2$, $r\dfrac{d^2\theta}{dt^2} + 2\dfrac{dr}{dt}\dfrac{d\theta}{dt}$

34.   Suppose a particle moves on the cardioid $r = a(1 + \cos \theta)$ and that the radius vector turns counterclockwise at the constant rate $\frac{d\theta}{dt} = \frac{5}{a\sqrt{3}}$ rad/sec. Let us find the polar form of the velocity and acceleration vectors when $\theta = \frac{\pi}{3}$. Now,

$\frac{dr}{dt} = \frac{dr}{d\theta} \cdot \frac{d\theta}{dt} = $ _____ $\cdot \frac{5}{a\sqrt{3}} = $ _____ and

$r \frac{d\theta}{dt} = $ _____. Evaluation of these components at

$\theta = \frac{\pi}{3}$ gives

$\vec{v} = $ _____ $\vec{u}_r + $ _____ $\vec{u}_\theta$, the velocity vector.

Next,

$$\frac{d^2r}{dt^2} = \frac{d\dot{r}}{d\theta} \frac{d\theta}{dt} = \text{_____} \cdot \frac{5}{a\sqrt{3}} = \text{_____}$$

so that

$\frac{d^2r}{dt^2} - r\left(\frac{d\theta}{dt}\right)^2 = -\frac{25}{6a} - (\text{_____}) \frac{25}{3a^2} = $ _____ when $\theta = \frac{\pi}{3}$.

Also for $\theta = \frac{\pi}{3}$,

$r \frac{d^2\theta}{dt^2} + 2 \frac{dr}{dt} \frac{d\theta}{dt} = \frac{3a}{2}(0) + 2(\text{_____}) \frac{5}{a\sqrt{3}} = $ _____.

Therefore,

$\vec{a} = [\text{_____}]\vec{u}_r + [\text{_____}]\vec{u}_\theta$, the acceleration.

35.   Consider a particle moving along the curve defined parametrically by $r = e^t$ and $\theta = t^2$. Then,

$\frac{dr}{dt} = $ _____, $\frac{d\theta}{dt} = $ _____; $\frac{d^2r}{dt^2} = $ _____, $\frac{d^2\theta}{dt^2} = $ _____.

The velocity at any time t is given by

$\vec{v} = $ _____ $\vec{u}_r + $ _____ $\vec{u}_\theta$;

and the acceleration is given by

$a = [e^t - \text{_____}]\vec{u}_r + [2e^t + \text{_____}]\vec{u}_\theta$

$= $ _____ $\vec{u}_r + $ _____ $\vec{u}_\theta$.

---

34. $-a \sin \theta$, $-\frac{5}{\sqrt{3}} \sin \theta$, $\frac{5}{\sqrt{3}}(1 + \cos \theta)$, $-\frac{5}{2}$, $\frac{5\sqrt{3}}{2}$, $-\frac{5}{\sqrt{3}} \cos \theta$, $-\frac{25}{3a} \cos \theta$, $\frac{3a}{2}$, $-\frac{50}{3a}$, $-\frac{5}{2}$, $\frac{-25}{a\sqrt{3}}$, $-\frac{50}{3a}$, $\frac{-25}{a\sqrt{3}}$

35. e^t, $2t$, e^t, 2, e^t, $2te^t$, $4t^2e^t$, $4te^t$, $e^t\left(1 - 4t^2\right)$, $2e^t(1 + 2t)$

CHAPTER 12   SELF-TEST

1. Find the first and second derivatives $\vec{f}'(t)$ and $\vec{f}''(t)$ of the following vector functions.

    (a) $\vec{f}(t) = (\ln t)\vec{i} - t^{-2}\vec{j} + e^{t^2}\vec{k}, \quad t > 0$

    (b) $\vec{f}(t) = (t \sin t)\vec{i} + (\tan^{-1} t)\vec{j} + e^{-t^2}\vec{k}$

2. Find the velocity and acceleration vectors, and the speed of a particle moving along a curve according to the position vector given as a function of time.

    (a) $\vec{r} = (3t^2 - 5)\vec{i} + (\ln t)\vec{j}, \quad t > 0$

    (b) $\vec{r} = e^{-2t}\vec{i} + 5 \cos 3t\,\vec{j} + 5 \sin 3t\,\vec{k}$

3. Find the cosine of the angle between the velocity and acceleration vectors at any time t in Problem 2(b).

4. A cannon whose muzzle is tilted at an angle of elevation of $30°$ shoots a ball with an initial velocity of 1200 m/sec.
    (a) Find parametric equations describing the motion of the cannonball.
    (b) How far from the cannon does the ball land, and what is its velocity and speed at the time of impact with the earth?
    (c) How high does the ball get and what is the total time the ball spends in the air?
    Assume that the mouth of the cannon is at ground level and that gravity is the only force acting on the cannonball.

5. Find the unit tangent vector, the principal normal, the curvature, and the length of the curve over the interval $1 \leq t \leq 2$, for the plane curve
    $$\vec{r} = (2 \ln t)\vec{i} - \left(\frac{1}{t} + t\right)\vec{j} \ .$$

6. Find an equation of the osculating circle when $t = 1$ for the curve in Problem 5.

7. Find the unit tangent vector $\vec{T}$, the principal normal vector $\vec{N}$, the curvature κ, and the unit binormal vector $\vec{B}$ for the space curve given by
    $$\vec{r} = (t + 1)\vec{i} - t^2\vec{j} + (1 - 2t)\vec{k} \ .$$

8. Establish the result
    $$\frac{d}{dt}|\vec{u}| = \frac{\vec{u} \cdot \vec{u}'}{|\vec{u}|}$$

    where $\vec{u}$ is a differentiable vector function of t for which $|\vec{u}|$ is never zero. Here $\vec{u}'$ means $\frac{d\vec{u}}{dt}$.

9. Find the velocity and acceleration vectors, the speed $\frac{ds}{dt}$, and the tangential and normal components of acceleration for the motion described by

$$\vec{r} = t\vec{i} + (\ln t)\vec{j}, \quad t > 0$$

10. Find the curvature and torsion of the path

$$\vec{r} = e^t \cos t\,\vec{i} + e^t \sin t\,\vec{j} + e^t\vec{k}$$

when $t = 0$.

11. A particle moves counterclockwise on the circle $r = b \cos \theta$, $b > 0$, with constant speed v_0. Find the acceleration vector in terms of $\vec{u}_r$ and $\vec{u}_\theta$.

SOLUTIONS TO CHAPTER 12 SELF-TEST

1. (a) $\vec{f}'(t) = t^{-1}\vec{i} + 2t^{-3}\vec{j} + 2te^{t^2}\vec{k}$

$\vec{f}''(t) = -t^{-2}\vec{i} - 6t^{-4}\vec{j} + 2e^{t^2}(1 + 2t^2)\vec{k}$

(b) $\vec{f}'(t) = (\sin t + t \cos t)\vec{i} + \dfrac{1}{1 + t^2}\vec{j} - 2te^{-t^2}\vec{k}$

$\vec{f}''(t) = (2 \cos t - t \sin t)\vec{i} + \dfrac{-2t}{\left(1 + t^2\right)^2}\vec{j} + 2e^{-t^2}(2t^2 - 1)\vec{k}$

2. (a) $\vec{v} = \dfrac{d\vec{r}}{dt} = 6t\vec{i} + t^{-1}\vec{j}$, velocity

$\vec{a} = \dfrac{d\vec{v}}{dt} = 6\vec{i} - t^{-2}\vec{j}$, acceleration

$\dfrac{ds}{dt} = |\vec{v}| = \sqrt{36t^2 + t^{-2}}$, speed

(b) $\vec{v} = -2e^{-2t}\vec{i} - 15 \sin 3t\vec{j} + 15 \cos 3t\vec{k}$, velocity

$\vec{a} = \dfrac{d\vec{v}}{dt} = 4e^{-2t}\vec{i} - 45 \cos 3t\vec{j} - 45 \sin 3t\vec{k}$, acceleration

$\dfrac{ds}{dt} = |\vec{v}| = \sqrt{4e^{-4t} + 225}$, speed

3. $\cos \theta = \dfrac{\vec{a} \cdot \vec{v}}{|\vec{a}||\vec{v}|}$; $\vec{a} \cdot \vec{v} = -8e^{-4t}$ and $|\vec{a}| = \sqrt{16e^{-4t} + 2025}$.

Therefore, $\cos \theta = \dfrac{-8e^{-4t}}{\sqrt{4e^{-4t} + 225} \cdot \sqrt{16e^{-4t} + 2025}}$.

4. (a) $\dfrac{d^2x}{dt^2} = 0$ m/sec^2, $\dfrac{dx}{dt} = c_1$ m/sec, $x = c_1 t + c_2$ m,

$\dfrac{d^2y}{dt^2} = -g$ m/sec^2, $\dfrac{dy}{dt} = -gt + c_3$ m/sec,

$y = -\frac{1}{2}gt^2 + c_3 t + c_4$ m. From the initial conditions, $t = 0$, $x = 0$, $y = 0$, $\dfrac{dx}{dt}(0) = v_0 \cos 30° = 600\sqrt{3}$ and $\dfrac{dy}{dt}(0) = v_0 \sin 30° = 600$, we find $c_1 = 600\sqrt{3}$, $c_2 = 0$, $c_3 = 600$, and $c_4 = 0$. Thus,

$$x = 600\sqrt{3}t \quad \text{and} \quad y = -\frac{1}{2}gt^2 + 600t$$

are parametric equations for the motion. Here, $g \approx 9.81$ m/sec^2.

(b) The maximum range occurs when $y = 0$ or $t = \dfrac{2 \cdot 600}{g}$ ≈ 122.3 sec. The range is then $R = 600\sqrt{3}\left(\dfrac{1200}{g}\right)$ $\approx 127{,}123$ m $= 1{,}271.23$ km. The velocity at any time is the vector $\vec{v} = \dfrac{dx}{dt}\vec{i} + \dfrac{dy}{dt}\vec{j}$. When $t = \dfrac{1200}{g}$, we find

$$\vec{v} = 600\sqrt{3}\,\vec{i} + \left(600 - g \cdot \frac{1200}{g}\right)\vec{j} = 600(\sqrt{3}\,\vec{i} - \vec{j}) ,$$

as the velocity of impact. Then, $|\vec{v}| = 600\sqrt{4} \approx 1200$ m/sec is the speed on impact with the earth.

(c)   The maximum height of the cannonball occurs when $\frac{dy}{dt} = 0$ or $t_m = \frac{600}{g}$;   then

$$h = - \tfrac{1}{2}g\left(\frac{600}{g}\right)^2 + 600\left(\frac{600}{g}\right) \approx 183.49 \text{ km}$$

is the maximum height attained. Finally, the ball remains in the air

$$2t_m = 2\left(\frac{600}{g}\right) \approx 122.3 \text{ sec .}$$

5. $\vec{v} = \frac{d\vec{r}}{dt} = 2t^{-1}\vec{i} + \frac{1 - t^2}{t^2}\vec{j}$ so that

$$|\vec{v}| = \sqrt{\frac{4}{t^2} + \frac{1 - 2t^2 + t^4}{t^4}} = \frac{t^2 + 1}{t^2} .$$

Thus the unit tangent vector is

$$\vec{T} = \frac{\vec{v}}{|\vec{v}|} = \frac{2t}{1 + t^2}\vec{i} + \frac{1 - t^2}{1 + t^2}\vec{j}$$

and the principal normal is

$$\vec{N} = - \frac{1 - t^2}{1 + t^2}\vec{i} + \frac{2t}{1 + t^2}\vec{j}.$$

The length of the curve is given by the integral

$$s = \int_1^2 |\vec{v}|\,dt = \int_1^2 \left(1 + t^{-2}\right)dt = t - \tfrac{1}{t}\Big|_1^2 = \tfrac{3}{2}.$$

To calculate the curvature,

$$\dot{x} = \tfrac{2}{t}, \quad \dot{y} = \tfrac{1}{t^2} - 1, \quad \ddot{x} = - \tfrac{2}{t^2}, \quad \ddot{y} = - \tfrac{2}{t^3}$$

gives

$$\kappa = \frac{|\dot{x}\ddot{y} - \dot{y}\ddot{x}|}{|\vec{v}|^3} = \frac{|-4t^{-4} + 2t^{-4} - 2t^{-2}|}{\left(t^2 + 1\right)^3 t^{-6}} = \frac{2t^2}{\left(t^2 + 1\right)^2} .$$

6.   From our calculation in the solution to Problem 5,

$$\rho = \tfrac{1}{\kappa} = \frac{\left(t^2 + 1\right)^2}{2t^2}, \quad \text{so} \quad \rho = 2 \quad \text{is the radius of curvature when}$$

t = 1. Also, $\vec{N} = - \frac{1 - t^2}{1 + t^2}\vec{i} + \frac{2t}{1 + t^2}\vec{j}$ so that $\vec{N} = \vec{j}$, and

$\vec{r} = -2\vec{j}$ is the position of the particle at t = 1; i.e., the particle is located at the point (0,-2). The center is located $\rho = 2$ units from the position point (0,-2) in the direction $\vec{N} = \vec{j}$; thus, the center of the osculating circle is (0,0). Therefore,

$$x^2 + y^2 = 4$$

is an equation of the osculating circle when t = 1.

7. $\vec{v} = \dfrac{d\vec{r}}{dt} = \vec{i} - 2t\vec{j} - 2\vec{k}$, $|\vec{v}| = \dfrac{ds}{dt} = \sqrt{5 + 4t^2}$.   Thus,

$\vec{T} = \dfrac{\vec{v}}{|\vec{v}|} = \left(5 + 4t^2\right)^{-1/2}(\vec{i} - 2t\vec{j} - 2\vec{k})$.

$\dfrac{d\vec{T}}{dt} = -\dfrac{1}{2}\left(5 + 4t^2\right)^{-3/2}8t\,[\vec{i} - 2t\vec{j} - 2\vec{k}] + \left(5 + 4t^2\right)^{-1/2}(-2\vec{j})$

$\qquad = 2\left(5 + 4t^2\right)^{-3/2}[-2t\vec{i} - 5\vec{j} + 4t\vec{k}]$;

$\dfrac{d\vec{T}}{ds} = \dfrac{d\vec{T}/dt}{ds/dt} = 2\left(5 + 4t^2\right)^{-2}[-2t\vec{i} - 5\vec{j} + 4t\vec{k}]$;

$\kappa = \left|\dfrac{d\vec{T}}{ds}\right| = 2\left(5 + 4t^2\right)^{-2}\sqrt{20t^2 + 25} = 2\sqrt{5}\left(5 + 4t^2\right)^{-3/2}$;

$\vec{N} = \dfrac{1}{k}\dfrac{d\vec{T}}{ds} = \left(5 + 4t^2\right)^{-1/2}[-\dfrac{2t}{\sqrt{5}}\vec{i} + \sqrt{5}\vec{j} + \dfrac{4t}{\sqrt{5}}\vec{k}]$

$\vec{B} = \vec{T} \times \vec{N} = -\dfrac{2}{\sqrt{5}}\vec{i} - \dfrac{1}{\sqrt{5}}\vec{k}$.

8. Now, $|\vec{u}| = \sqrt{\vec{u} \cdot \vec{u}}$ so that

$\dfrac{d}{dt}|\vec{u}| = \dfrac{d}{dt}\sqrt{\vec{u} \cdot \vec{u}} = \dfrac{1}{2\sqrt{\vec{u} \cdot \vec{u}}}\dfrac{d}{dt}(\vec{u} \cdot \vec{u}) = \dfrac{\vec{u}' \cdot \vec{u} + \vec{u} \cdot \vec{u}'}{2\sqrt{\vec{u} \cdot \vec{u}}} = \dfrac{\vec{u} \cdot \vec{u}'}{|\vec{u}|}$,

as claimed.

9. $\vec{v} = \dfrac{d\vec{r}}{dt} = \vec{i} + t^{-1}\vec{j}$,   velocity vector;

$\vec{a} = \dfrac{d\vec{v}}{dt} = -t^{-2}\vec{j}$,   acceleration vector;

$|\vec{v}| = \dfrac{ds}{dt} = \left(1 + t^{-2}\right)^{1/2} = \dfrac{\sqrt{1 + t^2}}{t}$,   speed;

$\dfrac{d^2s}{dt^2} = -t^{-2}\sqrt{1 + t^2} + t^{-1}\cdot\dfrac{t}{\sqrt{1 + t^2}}$ or,

$a_T = \dfrac{-1}{t^2\sqrt{1 + t^2}}$,   tangential component of acceleration

$a_N = \sqrt{|\vec{a}|^2 - a_T^2} = \sqrt{\dfrac{1}{t^4} - \dfrac{1}{t^4(1 + t^2)}}$

$\qquad = \sqrt{\dfrac{t^2}{t^4(1 + t^2)}} = \dfrac{1}{t\sqrt{1 + t^2}}$,   normal component of acceleration

10. $\vec{v} = e^t(\cos t - \sin t)\vec{i} + e^t(\cos t + \sin t)\vec{j} + e^t\vec{k}$ and

$\vec{a} = -2e^t \sin t\,\vec{i} + 2e^t \cos t\,\vec{j} + e^t\vec{k}$

Thus, when $t = 0$, $\vec{v}(0) = \vec{i} + \vec{j} + \vec{k}$ and $\vec{a}(0) = 2\vec{j} + \vec{k}$.   Then,

$$\vec{v}(0) \times \vec{a}(0) = \begin{vmatrix} \vec{i} & \vec{j} & \vec{k} \\ 1 & 1 & 1 \\ 0 & 2 & 1 \end{vmatrix} = -\vec{i} - \vec{j} + 2\vec{k}$$

and $|\vec{v}(0) \times \vec{a}(0)| = \sqrt{6}$, $|\vec{v}(0)| = \sqrt{3}$. Therefore, the curvature is given by

$$\kappa = \frac{|\vec{v}(0) \times \vec{a}(0)|}{|\vec{v}(0)|^3} = \frac{\sqrt{6}}{3\sqrt{3}} = \frac{\sqrt{2}}{3}.$$

Using formula (16) on page 801 of Finney/Thomas, the torsion is given by

$$\tau = \pm \frac{\begin{vmatrix} e^t(\cos t - \sin t) & e^t(\sin t + \cos t) & e^t \\ -2e^t \sin t & 2e^t \cos t & e^t \\ -2e^t(\sin t + \cos t) & 2e^t(\cos t - \sin t) & e^t \end{vmatrix}}{|\vec{v}(0) \times \vec{a}(0)|^2}$$

evaluated at $t = 0$, or

$$\tau = \frac{\begin{vmatrix} 1 & 1 & 1 \\ 0 & 2 & 1 \\ -2 & 2 & 1 \end{vmatrix}}{(\sqrt{6})^2} = \frac{1}{3} \quad \text{when} \quad t = 0.$$

11. $\frac{dr}{dt} = -b \sin\theta \frac{d\theta}{dt}$ and $r \frac{d\theta}{dt} = b \cos\theta \frac{d\theta}{dt}$.

Since $\vec{v} = \frac{dr}{dt} \vec{u}_r + r \frac{d\theta}{dt} \vec{u}_\theta$, we have

$$|\vec{v}| = \vec{v}_0 = \sqrt{b^2 \sin^2\theta + b^2 \cos^2\theta} \left|\frac{d\theta}{dt}\right| = b \frac{d\theta}{dt},$$

since $b > 0$ and the particle is moving counterclockwise.

Therefore, $\frac{d\theta}{dt} = \frac{\vec{v}_0}{b}$ is constant. Next,

$$\frac{d^2r}{dt^2} = \frac{d\dot{r}}{d\theta} \frac{d\theta}{dt} = -b \cos\theta \left(\frac{d\theta}{dt}\right)^2 = -\frac{\vec{v}_0^2}{b} \cos\theta$$

so

$$\frac{d^2r}{dt^2} - r\left(\frac{d\theta}{dt}\right)^2 = -\frac{\vec{v}_0^2}{b} \cos\theta - (b \cos\theta)\left(\frac{\vec{v}_0}{b}\right)^2$$

$$= -\frac{2\vec{v}_0^2}{b} \cos\theta, \quad \vec{u}_r\text{-component of } \vec{a};$$

Now $\frac{d^2\theta}{dt^2} = 0$, and we find

$$2 \frac{dr}{dt} \frac{d\theta}{dt} = -2b \sin \theta \left(\frac{d\theta}{dt}\right)^2 = -\frac{2\vec{v}_0^2}{b} \sin \theta, \quad \vec{u}_\theta\text{-component of } \vec{a}.$$

Hence, the acceleration vector is given by

$$\vec{a} = \left(-\frac{2\vec{v}_0^2}{b} \cos \theta\right)\vec{u}_r + \left(-\frac{2\vec{v}_0^2}{b} \sin \theta\right)\vec{u}_\theta .$$

NOTES:

CHAPTER 13 FUNCTIONS OF TWO OR MORE VARIABLES AND THEIR DERIVATIVES

13.1 FUNCTIONS OF TWO OR MORE INDEPENDENT VARIABLES

OBJECTIVE A : Given a function $w = f(x,y)$, find its domain and range.

1. For the function $w = \sqrt{x^2 - y^2}$, in order that w be a real number, the expression $x^2 - y^2$ must be _____ or, $x^2 \geq$ _____. Equivalently, $|x| \geq$ _____. Sketch a graph of the domain set in the coordinate system at the right.

Since $y = 0$ gives $w = \sqrt{x^2}$, the <u>range</u> of f is the set of values _____.

2. Consider the function $w = \frac{x}{y}e^{-x^2}$. Now $\frac{x}{y}$ is a real number for any value of x and any value of $y \neq$ _____. Also, e^{-x^2} is a real number for x satisfying _____. Therefore, w is a real number except when _____; that is, the domain of w contains all points in the xy-plane except the _____.

OBJECTIVE B : Given a function $w = f(x,y)$, represent the function
(a) by sketching a surface in space, and
(b) by drawing a family of level curves in the plane.

3. Let $f(x,y) = 4x^2 + 9y^2$. The equation $w = 4x^2 + 9y^2$ describes an elliptic _____ in xyw-space. The section cut out from the surface by the yw-plane is $x = 0$, _____ which is a parabola with vertex at the origin, opening upward. When

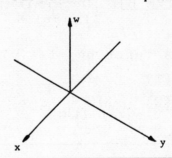

_____

1. nonnegative, y^2, $|y|$, shaded portion in figure gives (x,y) points where $|x| \geq |y|$, $w \geq 0$

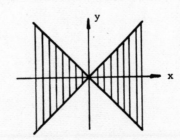

2. 0, $-\infty < x < \infty$, $y = 0$, x-axis

$y = 0$, _____ is also a parabola opening upward. Sketch the graph of the surface.

4.   A level curve for the constant $c > 0$, and function in Problem 3, is given by the equation

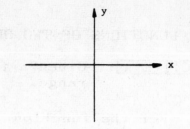

$c =$ _____ .

For a fixed value of c, the level curve is an _____ in the xy-plane. If $c = 0$ the level curve consists of a single point, namely the origin.

13.2   LIMITS AND CONTINUITY

OBJECTIVE A :   Given an elementary function of two variables x and y,   find its limit as (x,y) approches the point (a,b),   if the limit exists.

5. $\lim\limits_{(x,y)\to(0,0)} \dfrac{e^x + e^{-y}}{xy + 1} = \dfrac{1 + \rule{1cm}{0.1pt}}{0 + 1} = $ _____ .

6. $\lim\limits_{(x,y)\to(3,3)} \tan^{-1} \dfrac{y}{x} = \tan^{-1} \dfrac{3}{\rule{1cm}{0.1pt}} = \tan^{-1} \rule{1cm}{0.1pt} = $ _____ .

7.   In order for $f(x,y)$ to have a limit as $(x,y) \to (a,b)$,   the limits along <u>all</u> paths of approach have to _____ .

OBJECTIVE B :   Determine the points in the xy-plane at which a given function $f(x,y)$ is continuous.

8.   A function $f(x,y)$ is continuous at (a,b) in the plane if

      (1) f is _____ at (a,b),

      (2) $\lim\limits_{(x,y)\to(a,b)} f(x,y)$ _____ ,

---

3. paraboloid, $w = 9y^2$, $w = 4x^2$

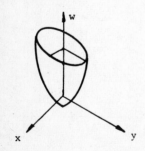

4. $4x^2 + 9y^2$, ellipse

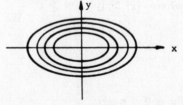

5. 1, 2   6. 3, 1, $\frac{\pi}{4}$ 7. agree

(3) $\lim_{(x,y)\to(a,b)} f(x,y) = $ _____ .

9. The function $f(x,y) = \cos(x - y)$ is defined at every point in the plane. Moreover,

$$\lim_{(x,y)\to(a,b)} \cos(x - y) = \text{_____}$$

and

$$f(a,b) = \text{_____} .$$

We conclude that $\cos(x - y)$ is continuous at every point in the plane.

10. The function $f(x,y) = \dfrac{x^2 - xy}{x - y}$ is defined and continuous at all points in the plane except along the line _____ .
However, for any point (a,a) in the plane,

$$\lim_{(x,y)\to(a,a)} \frac{x^2 - xy}{x - y} = \lim_{(x,y)\to(a,a)} \frac{x(x - y)}{x - y}$$

$$= \lim_{(x,y)\to(a,a)} \text{_____} , \quad \text{provided } x \neq y$$

$$= \text{_____} .$$

Therefore, the limit of $f(x,y)$ _does exist_ for every point in the plane. Nevertheless, since $f(x,y)$ fails to be defined when $x = y$, the function is discontinuous at every point (a,a) along the line $x = y$.

13.3 PARTIAL DERIVATIVES

OBJECTIVE A : Given an equation of a real-valued function of several variables, find the partial derivatives with respect to each variable.

11. $f(x,y) = xe^{-y} + y \cos x$
$f_x(x,y) = e^{-y} + $ _____ and $f_y(x,y) = $ _____ $+ \cos x$.

12. $w = xy e^{-y^2/2}$
$\dfrac{\partial w}{\partial x} = $ _____ and $\dfrac{\partial w}{\partial y} = xe^{-y^2/2} - $ _____ .

13. $g(x,y,z) = \dfrac{z}{y} + \tan(xy - 1)$
$g_x(x,y,z) = $ _____ ,
$g_y(x,y,z) = $ _____ $+ x \sec^2(xy - 1)$, and $g_z(x,y,z) = $ _____ .

_____

8. defined, exists, f(a,b)   9. $\cos(a - b)$, $\cos(a - b)$ 10. $y = x$, x, a   11. $-y \sin x$, $-xe^{-y}$

12. $ye^{-y^2/2}$, $xy^2 e^{-y^2/2}$ 13. $y \sec^2(xy - 1)$, $-\dfrac{z}{y^2}$, $\dfrac{1}{y}$

14. $h(x,y,z,w,\lambda,\mu) = (x - w)^2 + (y - z)^2 - \lambda(y - x - 1) - \mu(w - z^2)$

   $h_x(x,y,z,w,\lambda,\mu) = 2(x - w) +$ _____ ,

   $h_y(x,y,z,w,\lambda,\mu) =$ _____ $- \lambda$,

   $h_z(x,y,z,w,\lambda,\mu) = -2(y - z) +$ _____ ,

   $h_w(x,y,z,w,\lambda,\mu) =$ _____ $- \mu$,

   $h_\lambda(x,y,z,w,\lambda,\mu) =$ _____ , and

   $h_\mu(x,y,z,w,\lambda,\mu) =$ _____ .

OBJECTIVE B : Given a function of several independent variables, calculate all partial derivatives of the second-order.

15.   For the function $f(x,y) = xe^{-y} + y \cos x$ we found in Problem 11 above that,

$$f_x = e^{-y} - y \sin x \quad \text{and} \quad f_y = -xe^{-y} + \cos x.$$

Hence, the second-order partials are given by

$f_{xx} = \frac{\partial}{\partial x}(f_x) =$ _____ , $f_{xy} = \frac{\partial}{\partial y}(f_x) =$ _____ ,

$f_{yx} =$ _____ $= -e^{-y} - \sin x$, $f_{yy} =$ _____ $=$ _____ .

16.   Let $g(x,y,z) = x^3y + xyz + 3yz^2$. Then,

   $g_x = 3x^2y + yz$, $g_y =$ _____ , $g_z = xy + 6yz$.

   The second-order partials are given by

   $g_{xx} =$ _____ , $g_{xy} =$ _____ , $g_{xz} = y$,

   $g_{yx} = 3x^2 + z$, $g_{yy} =$ _____ , $g_{yz} =$ _____ ,

   $g_{zx} =$ _____ , $g_{zy} = x + 6z$, $g_{zz} =$ _____ .

   Note that the various mixed derivatives are equal: $g_{xy} = g_{yx}$, $g_{xz} = g_{zx}$, and $g_{yz} = g_{zy}$.

13.4   THE CHAIN RULE

OBJECTIVE A : Let w be a differentiable function of the variables $x,y,z,\ldots,v$ and let these in turn be differentiable functions of a second set of variables $p,q,r,\ldots,t$. Calculate the derivative of w with respect to any one of the variables in the second set by use of the chain rule for partial derivatives.

---

14. λ, $2(y - z)$, $2z\mu$, $-2(x - w)$, $-y + x + 1$, $-w + z^2$

15. $-y \cos x$, $-e^{-y} - \sin x$, $\frac{\partial}{\partial x}(f_y)$, $\frac{\partial}{\partial y}(f_y)$, xe^{-y}

16. $x^3 + xz + 3z^2$, $6xy$, $3x^2 + z$, 0, $x + 6z$, y, $6y$

17. To express $\frac{dw}{dt}$ as a function of t if $w = x^2 - xy$ and $x = e^t$, $y = \ln t$ we first differentiate w with respect to each of the variables x,y in the first set:

$$\frac{\partial w}{\partial x} = \underline{\hspace{2cm}} \quad \text{and} \quad \frac{\partial w}{\partial y} = \underline{\hspace{2cm}}.$$

Next, we differentiate each of the variables x,y with respect to the one variable t of the second set:

$$\frac{dx}{dt} = \underline{\hspace{2cm}} \quad \text{and} \quad \frac{dy}{dt} = \underline{\hspace{2cm}}.$$

Then we form the products of corresponding derivatives, add these products together, and substitute as follows:

$$\frac{dw}{dt} = \frac{\partial w}{\partial x} \frac{dx}{dt} + \underline{\hspace{2cm}} = (2x - y)e^t + \underline{\hspace{2cm}}$$
$$= (2e^t - \ln t)e^t - \underline{\hspace{2cm}}.$$

18. To express $\frac{\partial w}{\partial u}$ and $\frac{\partial w}{\partial v}$ as functions of u and v if $w = x^y + z$ and $x = 1 + u^2$, $y = e^v$, $z = uv$ we first differentiate w with respect to each of the variables x,y,z in the first set:

$$\frac{\partial w}{\partial x} = \underline{\hspace{1.5cm}}, \quad \frac{\partial w}{\partial y} = \underline{\hspace{1.5cm}}, \quad \frac{\partial w}{\partial z} = \underline{\hspace{1.5cm}}.$$

Next, we differentiate each of those variables with respect to the first variable u of the second set:

$$\frac{\partial x}{\partial u} = \underline{\hspace{1cm}}, \quad \frac{\partial y}{\partial u} = 0, \quad \frac{\partial z}{\partial u} = \underline{\hspace{1cm}}.$$

Then

$$\frac{\partial w}{\partial u} = \frac{\partial w}{\partial x} \frac{\partial x}{\partial u} + \frac{\partial w}{\partial y} \frac{\partial y}{\partial u} + \underline{\hspace{2cm}}$$
$$= (yx^{y-1})(2u) + (x^y \ln x)(0) + \underline{\hspace{1cm}} = e^v(1 + u^2)^{e^v-1}(2u) + v.$$

To calculate $\frac{\partial w}{\partial v}$ we differentiate each of the variables x,y,z in the first set with respect to the second variable v of the second set:

$$\frac{\partial x}{\partial v} = 0, \quad \frac{\partial y}{\partial v} = \underline{\hspace{1.5cm}}, \quad \frac{\partial z}{\partial v} = \underline{\hspace{1.5cm}}.$$

Then,

$$\frac{\partial w}{\partial v} = \frac{\partial w}{\partial x} \frac{\partial x}{\partial v} + \underline{\hspace{2.5cm}}$$
$$= (yx^{y-1})(0) + \underline{\hspace{2cm}} + (1)(u) = \underline{\hspace{3cm}}.$$

---

17. $2x - y$, $-x$, e^t, $\frac{1}{t}$, $\frac{\partial w}{\partial y} \frac{dy}{dt}$, $(-x)\frac{1}{t}$, $\frac{1}{t}e^t$

18. yx^{y-1}, $x^y \ln x$, 1, $2u$, v, $\frac{\partial w}{\partial z} \frac{\partial z}{\partial u}$, $1(v)$, e^v, u, $\frac{\partial w}{\partial y} \frac{\partial y}{\partial v} + \frac{\partial w}{\partial z} \frac{\partial z}{\partial v}$, $(x^y \ln x)e^v$,

$(1 + u^2)^{e^v} \ln(1 + u^2)e^v + u$

19.   Suppose that $w = f\left(\frac{y}{x}\right)$ for a differentiable function f. Show

that $\frac{\partial w}{\partial x} = -\frac{y}{x^2} f'\left(\frac{y}{x}\right)$ and $\frac{\partial w}{\partial y} = \frac{1}{x} f'\left(\frac{y}{x}\right)$.

Solution. Let $w = f(u)$ where $u = \frac{y}{x}$. To find the partial
derivatives of w, we first differentiate w with respect to
the single variable u of the first set: $\frac{dw}{du} = f'(u)$. To find
$\frac{\partial w}{\partial x}$ we next calculate the derivative of the variable u from
the first set with respect to the first variable x of the
second set: $\frac{\partial u}{\partial x} = $ _____. Then the chain rule gives

$$\frac{\partial w}{\partial x} = \frac{dw}{du} \frac{\partial u}{\partial x} = \text{_____} \text{ as claimed.}$$

To find $\frac{\partial w}{\partial y}$ we calculate the derivative of the variable u
from the first set with respect to the second variable y of
the second set: $\frac{\partial u}{\partial y} = $ _____. Then the chain rule gives

$$\frac{\partial w}{\partial y} = \text{_____} = \text{_____} .$$

20.   A point moves on the surface $w = x^3 + 2xy^2$ with x increasing
at the rate of 4 cm/sec and y increasing at 6 cm/sec.
At what rate is $\frac{\partial w}{\partial x}$ changing at the moment when $x = 2$ cm
and $y = 5$ cm?

Solution. Let $u = \frac{\partial w}{\partial x} = $ _____. We want to calculate $\frac{du}{dt}$.
Now, by the chain rule,

$$\frac{du}{dt} = \text{_____} = (6x) \frac{dx}{dt} + \text{_____} .$$

At the specified moment,

$$\frac{du}{dt} = (12)(4) + \text{_____} = \text{_____} \text{ cm}^2/\text{sec.}$$

OBJECTIVE B : Assuming a given equation $F(x,y,z) = 0$ determines z
as a differentiable function of x and y, calculate
the partial derivatives $\frac{\partial z}{\partial x}$ and $\frac{\partial z}{\partial y}$ at points where
$F_z \neq 0$.

21.   Consider the equation $xz^2 + xy - y^3 = 0$. Denote the left side
in functional notation by $F(x,y,z)$. Then

$F_x = $ _____, $F_y = $ _____,   and $F_z = $ _____.

---

19. $-\dfrac{y}{x^2}$, $f'(u)\left(-\dfrac{y}{x^2}\right)$, $\dfrac{1}{x}$, $\dfrac{dw}{du}\dfrac{\partial u}{\partial y}$, $f'(u)\left(\dfrac{1}{x}\right)$

20. $3x^2 + 2y^2$, $\dfrac{\partial u}{\partial x}\dfrac{dx}{dt} + \dfrac{\partial u}{\partial y}\dfrac{dy}{dt}$, $(4y)\dfrac{dy}{dt}$, $(20)(6)$, 168

Thus, at all points where $2xz^2 \neq 0$,

$$\frac{\partial z}{\partial x} = \frac{}{2xz^2} \quad \text{and} \quad \frac{\partial z}{\partial y} = \frac{}{2xz^2} \ .$$

13.5   DIRECTIONAL DERIVATIVES AND GRADIENT VECTORS

OBJECTIVE A : Given a function $w = f(x,y,z)$, find the gradient vector at a specified point $P_0(x_0,y_0,z_0)$.

22.   For the function $f(x,y,z) = y \ln x^2 + x \cos z$ and the point $P_0(-1, 2, \frac{\pi}{2})$, the gradient vector is obtained by calculating the partial derivatives of f at P_0:

$f_x = $ _____ $|_0 = -4$, $f_y = \ln x^2 |_0 = $ _____ ,

$f_z = $ _____ $|_0 = $ _____ .   Thus, $\vec{\nabla} f = $ _____ .

OBJECTIVE B : Given a function f of two or three variables, find the directional derivative of f at a given point, and in the direction of a given vector $\vec{A}$.

23.   Let us find the directional derivative of $f = e^{x \tan y}$ at the point $(1, \frac{\pi}{4})$, and in the direction $\vec{A} = \frac{1}{e}\vec{i} + \frac{2}{e}\vec{j}$. Now,

$$\frac{\partial f}{\partial x} = \text{_____} \quad \text{and} \quad \frac{\partial f}{\partial y} = \text{_____} .$$

Thus, we obtain the gradient vector

$$\vec{\nabla} f = f_x\left(1, \frac{\pi}{4}\right)\vec{i} + f_y\left(1, \frac{\pi}{4}\right)\vec{j} = \text{_____} .$$

Next, $|\vec{A}| = $ _____ ,   and hence a unit vector in the direction of $\vec{A}$ is given by

$$\vec{u} = \frac{\vec{A}}{|\vec{A}|} = \text{_____} .$$

The derivative of f in the direction $\vec{A}$ can now be calculated from the vectors $\vec{u}$ and grad f as

$$\vec{u} \cdot \vec{\nabla} f = \frac{1}{\sqrt{5}}(\text{_____}) + \text{_____} (2e) = \text{_____} .$$

---

21. y, $x - 3y^2$, $2xz^2$, $-y$, $-(x - 3y^2)$

22. $\frac{2y}{x} + \cos z$, $\ 0$, $\ -x \sin z$, $\ 1$, $\ -4\vec{i} + 0\vec{j} + \vec{k}$

23. $\tan y \ e^{x \tan y}$, $\ x \sec^2 y \ e^{x \tan y}$, $\ e\vec{i} + 2e\vec{j}$, $\ \frac{\sqrt{5}}{e}$, $\ \frac{1}{\sqrt{5}}\vec{i} + \frac{2}{\sqrt{5}}\vec{j}$, $\ e$, $\ \frac{2}{\sqrt{5}}$, $\ \sqrt{5}e$

24. Consider the function $f = x^2 + 5y \sin z$. To find the directional derivative of f at the point $P_0(1, 1, \frac{\pi}{2})$ in the direction from P_0 toward the point $P_1(4, 5, \frac{\pi}{2})$, we first calculate the partial derivatives of f at P_0. Thus,

$f_x = $ _____$|_{(1,1,\pi/2)} = $ _____, $f_y = 5 \sin z|_{(1,1,\pi/2)} = $ _____,

$f_z = $ _____$|_{(1,1,\pi/2)} = 0$.

Therefore, the gradient of f at $(1, 1, \frac{\pi}{2})$ is

$$\vec{\nabla} f = \underline{\hspace{3cm}}.$$

The vector $\vec{A} = \vec{P_0 P_1} = 3\vec{i} + 4\vec{j} + 0\vec{k}$ has magnitude

$$|\vec{A}| = \sqrt{\underline{\hspace{3cm}}} = \underline{\hspace{1cm}}.$$

Thus, the direction of $\vec{A}$ is

$$\vec{u} = \frac{\vec{A}}{|\vec{A}|} = \underline{\hspace{4cm}}.$$

The derivative of f in the direction $\vec{A}$ is

$$\vec{u} \cdot \vec{\nabla} f = \frac{3}{5}(\underline{\hspace{1cm}}) + \underline{\hspace{1cm}} (5) + 0(0) = \underline{\hspace{2cm}}.$$

OBJECTIVE C :   Given a function f of two or three variables, determine the direction one should travel, starting from a given point P_0, to obtain the most rapid rate of increase or decrease (whichever is specified) of the function.

25. The function $w = f(x, y, z)$ changes most rapidly from the point $P_0(x_0, y_0, z_0)$ in the direction of the vector _____. Moreover, the directional derivative in this direction equals the _____ of the gradient vector. In this case the gradient vector lies in 3-space, which contains the domain of the function f.

26. In which direction should one travel, starting from the point $P_0(1, 1, \frac{\pi}{2})$, in order to obtain the most rapid rate of decrease of the function $f = x^2 + 5y \sin z$? What is the instantaneous rate of change of f per unit of distance in this direction?

Solution. In Problem 24 we calculated $\vec{\nabla} f(P_0) = $ _____. Since the directional derivative of f in the direction $\vec{u}$ is given by

$$\vec{u} \cdot \vec{\nabla} f = |\vec{\nabla} f| \cos \theta ,$$

_____

24. $2x$, 2, 5, $5y \cos z$, $2\vec{i} + 5\vec{j} + 0\vec{k}$, $3^2 + 4^2 + 0^2$, 5, $\frac{3}{5}\vec{i} + \frac{4}{5}\vec{j} + 0\vec{k}$, 2, $\frac{4}{5}$, $\frac{26}{5}$

25. $\vec{\nabla} f$, magnitude

where θ is the angle between the unit vector $\vec{u}$ and the vector _____, the value of the directional derivative is smallest when $\cos \theta =$ _____ or $\theta =$ _____. Thus, $\vec{u} =$ _____. The value of the directional derivative in this direction is

$$\vec{u} \cdot \vec{\nabla}f = \underline{\hspace{2cm}} .$$

This value is the negative of the magnitude of the gradient.

27. The function $w = f(x,y)$ changes most rapidly from the point $P_0(x_0,y_0)$ in the direction of the gradient vector

$$\vec{\nabla}f = \underline{\hspace{4cm}}$$

in the xy-plane (which contains the domain of the function). The directional derivative in this direction equals the magnitude of the vector _____.

28. Consider the function $f = x^y$. At the point $P_0(e,0)$ the partial derivatives of f are

$$f_x = \underline{\hspace{1.5cm}}\Big|_{(e,0)} = \underline{\hspace{1.5cm}}, \quad f_y = \underline{\hspace{1.5cm}}\Big|_{(e,0)} = \underline{\hspace{1.5cm}}$$

so that $\vec{\nabla}f =$ _____. The maximum value of the directional derivative at P_0 occurs in the direction $\vec{u} =$ _____, and this value equals $|\vec{\nabla}f| =$ _____.

Remark. Notice that the gradient vector for a function of two variables lies in the xy-plane, whereas for a function of three variables it lies in space.

13.6 TANGENT PLANES AND NORMAL LINES

OBJECTIVE A: Find the plane which is tangent to the level surface $f(x,y,z) = $ constant at a specified point $P_0(x_0,y_0,z_0)$.

29. The gradient vector $\vec{\nabla}f(P_0)$ is _____ to the level surface

$$f(x,y,z) = \text{constant}$$

at the point $P_0(x_0,y_0,z_0)$ on the surface.

30. Consider the surface $z = e^{3x} \sin 3y + 2$. Let

$$w = f(x,y,z) = e^{3x} \sin 3y - z$$

so the equation of the surface has the form

$$f(x,y,z) = \text{constant}.$$

---

26. $2\vec{i} + 5\vec{j} + 0\vec{k}$, $\vec{\nabla}f$, -1, π, $-\frac{2}{\sqrt{29}}\vec{i} - \frac{5}{\sqrt{29}}\vec{j} + 0\vec{k}$, $-\sqrt{29}$ 27. $f_x(x_0,y_0)\vec{i} + f_y(x_0,y_0)\vec{j}$, $\vec{\nabla}f$

28. yx^{y-1}, 0, $x^y \ln x$, 1, $0\vec{i} + 1\vec{j}$, $\vec{j}$, 1   29. normal

In this case the constant is equal to _____. The point
$P_0(0, \frac{\pi}{6}, 3)$ lies on the surface and the vector

$$\vec{\nabla}f(P_0) = \left(f_x \vec{i} + f_y \vec{j} + f_z \vec{k}\right)_0$$

$$= \left(3e^{3x} \sin 3y \, \vec{i} + \underline{\hspace{3cm}} \vec{j} - \vec{k}\right)_0$$

$$= \underline{\hspace{3cm}}$$

is normal to the surface at P_0. Thus, an equation of the
tangent plane to the surface at P_0 is

_____ (x - 0) + 0(_____) + _____ (z - 3) = 0, or z = _____.

31. Since the gradient vector $3\vec{i} + 0\vec{j} - \vec{k}$ is normal to the
surface in Problem 30 at the point $P_0(0, \frac{\pi}{6}, 3)$, equations of
the normal to the surface at P_0 are

x = 0 + 3t, y = _____, and z = _____.

These equations also may be written in the form $\frac{x}{3}$ = _____,
y = _____.

OBJECTIVE B : Given a surface w = f(x,y), find an equation of the
tangent plane to the surface at a specified point P_0,
if the tangent plane exists.

32. The point $P_0(1,-2,3)$ lies on the surface $w = \sqrt{3x^2 - xy + y^2}$.
To find an equation of the tangent plane, calculate

$$f_x(1,-2) = \frac{1}{2}(6x - y)\left(3x^2 - xy + y\right)\Big|_{(1,-2)} = \underline{\hspace{2cm}};$$

$$f_y(1,-2) = \underline{\hspace{5cm}}\Big|_{(1,-2)} = -\frac{5}{6};$$

Therefore, an equation of the tangent plane is

$$\frac{4}{3}(x - 1) + \underline{\hspace{2cm}}(y + 2) + \underline{\hspace{2cm}}(w - 3) = 0$$

or,

$$8x - 5y - 6w = \underline{\hspace{2cm}}.$$

OBJECTIVE C : Given a surface w = f(x,y), find the normal line
(if it exists) to the surface at a specified point P_0.

---

30. -2, $3e^{3x} \cos 3y$, $3\vec{i} + 0\vec{j} - \vec{k}$, 3, $y - \frac{\pi}{6}$, -1, 3(x + 1)

31. $\frac{\pi}{6} - 0t$, 3 - t, $\frac{z-3}{-1}$, $\frac{\pi}{6}$

32. $\frac{4}{3}$, $\frac{1}{2}(2y - x)\left(3x^2 - xy + y^2\right)^{-1/2}$, $-\frac{5}{6}$, -1, 0

33. For the surface and point in Problem 32 above, for a normal vector $\vec{N}$ to the surface P_0 we may take $\vec{N} = 8\vec{i} +$ _____.
Thus, equations for the normal line are given by

$$x = \underline{\qquad} + 8t, \quad y = -2 - \underline{\qquad}, \quad w = \underline{\qquad}.$$

Alternatively, the normal line is

$$\frac{x - 1}{8} = \frac{y + 2}{\underline{\qquad}} = \underline{\qquad}.$$

OBJECTIVE D : Given two surfaces $f(x,y,z) = $ constant and $g(x,y,z) = $ constant, find parametric equations for the line tangent to the curve C of intersection of the surfaces at a specified point $P_0(x_0, y_0, z_0)$ on C.

34. Consider the surfaces $f(x,y,z) = x^2 + y - z = 1$ and $g(x,y,z) = x + y + z = 3$. The point $(1,1,1)$ lies on the curve of intersection of the two surfaces. The tangent line is _____ to both $\vec{\nabla}f$ and $\vec{\nabla}g$ at $(1,1,1)$. The vector $\vec{v} = $ _____ will therefore be parallel to the line. Now,

$$\vec{\nabla}f(1,1,1) = (2x\vec{i} + \vec{j} - \vec{k})_{(1,1,1)} = \underline{\qquad}$$

and

$$\vec{\nabla}g(1,1,1) = (\underline{\qquad})_{(1,1,1)} = \underline{\qquad}.$$

Hence,

$$\vec{v} = (\vec{i} + \vec{j} - \vec{k}) \times (\vec{i} + \vec{j} + \vec{k}) = \underline{\qquad}.$$

The tangent line is

$$x = 1 + 2t, \quad y = \underline{\qquad}, \quad z = \underline{\qquad}.$$

13.7 LINEARIZATION AND DIFFERENTIALS

OBJECTIVE A : Given a function $f(x,y)$, find the standard linear approximation to it near a specified point.

35. Given the function $f(x,y)$ the linearization of it at the point (x_0, y_0) is

$$L(x,y) = \underline{\qquad\qquad\qquad}.$$

The right side of the previous equation represents the _____ that is tangent to the surface $w = f(x,y)$ at the point (x_0, y_0).

---

33. $-5\vec{j} - 6\vec{k}$, 1, 5t, 3 - 6t, -5, $\frac{w - 3}{-6}$

34. perpendicular, $\vec{\nabla}f \times \vec{\nabla}g$, $\vec{i} + \vec{j} - \vec{k}$, $\vec{i} + \vec{j} + \vec{k}$, $\vec{i} + \vec{j} + \vec{k}$, $2\vec{i} - 2\vec{j}$, 1 - 2t, 1

35. $f(x_0, y_0) + f_x(x_0, y_0)(x - x_0) + f_y(x_0, y_0)(y - y_0)$, plane

36. For the surface $f(x,y) = 3xy + ye^x - x \sin y$ we have
$f_x(x,y) = $ _____ and $f_y(x,y) = 3x + e^x - x \cos y$.
Hence, near the point $(0, \frac{\pi}{2})$ the surface $w = f(x,y)$ can be
approximated linearly by the tangent plane
$L(x,y) = $ ____ + _____ $(x - 0) + (1)$ _____ = _____ .

$\boxed{\text{OBJECTIVE B}}$: Given a surface $w = f(x,y)$, find the change df
along the plane tangent to the surface at some given
point (x_0, y_0, w_0) for specified increments Δx and
Δy.

37. Consider the surface $w = f(x,y) = y \cos(x + y)$. For the point
$(\frac{\pi}{2}, \frac{\pi}{2}, \underline{\quad})$, the change along the tangent plane is given by

$$df = \Delta L = f_x(\tfrac{\pi}{2}, \tfrac{\pi}{2}) \Delta x + \underline{\qquad\qquad} .$$

Now, $\dfrac{\partial f}{\partial x} = $ _____ and $\dfrac{\partial f}{\partial y} = $ _____ .
Thus, $f_x(\frac{\pi}{2}, \frac{\pi}{2}) = 0$ and $f_y(\frac{\pi}{2}, \frac{\pi}{2}) = $ ____ . For the increments
$\Delta x = -0.1$ and $\Delta y = 0.05$ we have $df = \Delta L = $ _____ as the
change along the tangent plane. On the other hand, the actual
change along the surface itself is
$\Delta f = f(\frac{\pi}{2} - 0.1, \frac{\pi}{2} + 0.05) - f(\frac{\pi}{2}, \frac{\pi}{2})$
    $\approx f(1.471, 1.621) + $ ____ $\approx$ _____ $\cos(3.092) + 1.571$
    $\approx 1.621(-0.999) + 1.571 \approx -0.048$
with the calculations done on a calculator.

$\boxed{\text{OBJECTIVE C}}$: Use the approximation

$$\Delta f = f(x,y) - f(x_0, y_0) \approx \underbrace{f_x(x_0, y_0) \Delta x + f_y(x_0, y_0) \Delta y}_{df}$$

to discuss the sensitivity of $f(x,y)$ to small changes
in x and y near a given point (x_0, y_0).

38. The surface area of a cone of radius r and height h is
given by the formula

$$S = \pi r \sqrt{r^2 + h^2} .$$

To calculate ΔL, we determine

$\dfrac{\partial S}{\partial r} = \pi \sqrt{r^2 + h^2} + $ _____ , and $\dfrac{\partial S}{\partial h} = $ _____ .

---

36. $3y + ye^x - \sin y$, $\frac{\pi}{2}$, $(2\pi - 1)$, $y - \frac{\pi}{2}$, $(2\pi - 1)x + y$

37. $-\frac{\pi}{2}$, $f_y(\frac{\pi}{2}, \frac{\pi}{2}) \Delta y$, $-y \sin(x + y)$, $\cos(x + y) - y \sin(x + y)$, -1, -0.05, $\frac{\pi}{2}$, 1.621

Therefore, near the point $(r_0, h_0) = (3,4)$,

$$\Delta S \approx dS = \frac{\partial S}{\partial r}(r_0, h_0)\Delta r + \frac{\partial S}{\partial h}(r_0, h_0)\Delta h = \underline{\hspace{1cm}} \Delta r + \underline{\hspace{1cm}} \Delta h.$$

Hence, a one-unit change in r will change S by about $\frac{34\pi}{5}$ units. A one-unit change in h will change S by about $\underline{\hspace{1cm}}$ units. The surface area of a cone of radius 3 and height 4 is nearly $\underline{\hspace{1cm}}$ times more sensitive to a small change in r than it is to a change of the same size in h.

13.8   MAXIMA, MINIMA, AND SADDLE POINTS

OBJECTIVE A :   Given the surface $z = f(x,y)$ defined by a function f which has continuous partial derivatives over some region R, examine the surface for local extrema. Use the second derivative test to classify the relative extrema as a local maximum, local minimum, or saddle point.

39.   A necessary condition that must be satisfied if $z = f(x,y)$ has a maximum (or minimum) value occurring at the point (a,b) that is not on the boundary of the region R is that
$\underline{\hspace{7cm}}$.

40.   Assume (a,b) is an interior point of the region R.   Suppose $f_x(a,b) = f_y(a,b) = 0$ and $f_{xx}(a,b) \neq 0$. Then at (a,b) the function $f(x,y)$ has:
   (i) A local minimum if $\underline{\hspace{4cm}}$.
   (ii) A local maximum if $\underline{\hspace{4cm}}$.
   (iii) A saddle point if $\underline{\hspace{3cm}}$.
   (iv) The test is inconclusive if $\underline{\hspace{3.5cm}}$.

41.   Consider the function $f(x,y) = x^2 - xy + y^3 - x$. We set
$$\frac{\partial f}{\partial x} = 2x - y - 1 = 0 \quad \text{and} \quad \frac{\partial f}{\partial y} = \underline{\hspace{2cm}} = 0 .$$
The first equation is equivalent to $y = 2x - 1$, and substitution of this into the second equation gives $-x + 3(2x - 1)^2 = 0$, or $12x^2 - 13x + 13 = 0$. Thus, $x = \frac{1}{3}$ or $x = \frac{3}{4}$. It follows that the points $\underline{\hspace{2cm}}$ and $\underline{\hspace{2cm}}$ are critical points where f_x and f_y are simultaneously 0. Next we find, $f_{xx} = \underline{\hspace{1cm}}$, $f_{xy} = \underline{\hspace{1cm}}$, and $f_{yy} = 6y$. Let us test the critical point $(\frac{1}{3}, -\frac{1}{3})$. If $f_{xx}(\frac{1}{3}, -\frac{1}{3}) = \underline{\hspace{1cm}}$,

---

38. $\dfrac{\pi r^2}{\sqrt{r^2 + h^2}}$, $\dfrac{\pi r h}{\sqrt{r^2 + h^2}}$, $\dfrac{34\pi}{5}$, $\dfrac{12\pi}{5}$, $\dfrac{12\pi}{5}$, three

39. $\dfrac{\partial f}{\partial x} = 0$ and $\dfrac{\partial f}{\partial y} = 0$ at (a,b)

40. $f_{xx} > 0$ and $f_{xx}f_{yy} - f_{xy}^2 > 0$ at (a,b), $f_{xx} < 0$ and $f_{xx}f_{yy} - f_{xy}^2 > 0$ at (a,b),

   $f_{xx}f_{yy} - f_{xy}^2 < 0$ at (a,b), $f_{xx}f_{yy} - f_{xy}^2 = 0$ at (a,b)

$f_{xy}\left(\frac{1}{3},-\frac{1}{3}\right) =$ _____, and $f_{yy}\left(\frac{1}{3},-\frac{1}{3}\right) =$ _____, then $f_{xx}f_{yy} - f_{xy}^2 =$ -5. We conclude that the surface has a _____ at $\left(\frac{1}{3},-\frac{1}{3}\right)$. To test the point $\left(\frac{3}{4},\frac{1}{2}\right)$, we have $f_{xx}\left(\frac{3}{4},\frac{1}{2}\right) = 2$, $f_{xy}\left(\frac{3}{4},\frac{1}{2}\right) =$ _____, $f_{yy}\left(\frac{3}{4},\frac{1}{2}\right) = 3$. Then, $f_{xx}f_{yy} - f_{xy}^2 =$ _____ and $f_{xx} > 0$. We conclude that the surface has a _____ at $\left(\frac{3}{4},\frac{1}{2}\right)$.

42.   For the function $f(x,y) = x^3 - 3x^2y^2 + y^4$, we find

$$\frac{\partial f}{\partial x} = \underline{\hspace{2cm}} = 3x(\underline{\hspace{2cm}}) \text{ and}$$

$$\frac{\partial f}{\partial y} = \underline{\hspace{2cm}} = 2y(\underline{\hspace{2cm}}).$$

From $\frac{\partial f}{\partial x} = 0$ and $\frac{\partial f}{\partial y} = 0$, $x = 0$ if and only if $y = 0$ also. Hence _____ is one critical point. If $x \neq 0$, then $\frac{\partial f}{\partial x} = 3x(x - 2y^2) = 0$ demands $x = 2y^2$. Since $y \neq 0$, $2y(2y^2 - 3x^2) = 0$ gives $2y^2 - 3x^2 = 0$, and substitution of $x = 2y^2$ into this last equation gives _____ $= 0$ or, $x(1 - 3x) = 0$. It follows that $x =$ _____. From $x = 2y^2$ we find that $\left(\frac{1}{3},\frac{1}{\sqrt{6}}\right)$ and _____ are also critical points of

f. Next we test these critical points. First, we compute the second partial derivatives: $f_{xx} = 6(x - y^2)$, $f_{xy} =$ _____, and $f_{yy} = 6(2y^2 - x^2)$.

(a)   The critical point $(0,0)$:

$f_{xx}(0,0) = 0$, $f_{xy}(0,0) = 0$, and $f_{yy}(0,0) = 0$. Therefore, the second derivative test yields <u>no information</u> concerning the surface at the point $(0,0)$. However, a little further analysis shows that the surface has a saddle point at $(0,0)$.

(b)   The critical point $\left(\frac{1}{3},\frac{1}{\sqrt{6}}\right)$:

$f_{xx}\left(\frac{1}{3},\frac{1}{\sqrt{6}}\right) =$ _____, $f_{xy}\left(\frac{1}{3},\frac{1}{\sqrt{6}}\right) =$ _____, and $f_{yy}\left(\frac{1}{3},\frac{1}{\sqrt{6}}\right) = \frac{4}{3}$. Then, $f_{xx}f_{yy} - f_{xy}^2 =$ _____ and we conclude that the surface has a _____ at $\left(\frac{1}{3},\frac{1}{\sqrt{6}}\right)$.

(c)   The critical point $\left(\frac{1}{3},-\frac{1}{\sqrt{6}}\right)$:

$f_{xx}\left(\frac{1}{3},-\frac{1}{\sqrt{6}}\right) = 1$, $f_{xy}\left(\frac{1}{3},-\frac{1}{\sqrt{6}}\right) = \frac{4}{\sqrt{6}}$, and $f_{yy}\left(\frac{1}{3},-\frac{1}{\sqrt{6}}\right) =$ _____.

41. $-x + 3y^2$, $\left(\frac{1}{3},-\frac{1}{3}\right)$, $\left(\frac{3}{4},\frac{1}{2}\right)$, 2, -1, 2, -1, -2, saddle point, -1, 5, **local minimum**

Thus, $f_{xx}f_{yy} - f_{xy}^2 =$ _____   and we conclude that the

surface has a   _____   at $\left(\frac{1}{3}, -\frac{1}{\sqrt{6}}\right)$.

OBJECTIVE B :   Find the absolute maxima and minima of a given function
$f(x,y)$ over a specified domain.

43.   Consider the function $f(x,y) = x^3 - xy + y^2$ on the closed
triangular plate bounded by the lines $x = 0$, $y = 1$, $y = x$
in the first quadrant.   First we consider the interior points
of the region.   For these we set

$$f_x = \text{_____} = 0$$
$$f_y = \text{_____} = 0.$$

Thus, the second equation gives $x = 2y$ and substitution into
the first equation yields $12y^2 - y = 0$.   Then, $y = 0$ or
$y =$ _____.   Hence the critical points are   _____   and
_____.   However, the point   _____   lies outside of the
region so it will not be considered.

Next we consider the boundary of the triangular region.

(i) $x = 0$: $f(0,y) = y^2$ may be regarded as a function of y
on the closed interval $0 \leq y \leq 1$.   Its minimum occurs
at $y = 0$ where $f(0,0) = 0$ and its maximum occurs at
$y =$ _____   where $f(0,1) =$ _____.

(ii) $y = 1$: $f(x,1) =$ _____   may be regarded as a
function of x on the closed interval $0 \leq x \leq 1$.   Its
derivative is $3x^2 - 1$,   which is zero when $x =$ _____
in the interval.   We now determine the value of $f(x,1)$
at this point and the endpoints of the interval:

$f\left(\frac{1}{\sqrt{3}}, 1\right) =$ _____, $f(0,1) =$ ____, $f(1,1) =$ ____.

(iii) $y = x$: $f(x,x) =$ ____   may be regarded as a function of
x on the closed interval $0 \leq x \leq 1$.   Its derivative is
$3x^2$,   which is positive over the interval, so the
function is increasing.   Thus the minimum value occurs
at the endpoint   _____   and the maximum value occurs
at   _____.

Putting all of this information together, we find
that the function has an absolute maximum value of   1
occurring at $(0,1)$ and $(1,1)$ and an absolute minimum
value of   ____   occurring at   _____.

42. $3x^2 - 6xy^2$, $x - 2y^2$, $4y^3 - 6x^2y$, $2y^2 - 3x^2$, $(0,0)$, $x - 3x^2$, $\frac{1}{3}$, $\left(\frac{1}{3}, -\frac{1}{\sqrt{6}}\right)$, $-12xy$,   1,
$-\frac{4}{\sqrt{6}}$, $-\frac{4}{3}$,   saddle point, $\frac{4}{3}$, $-\frac{4}{3}$,   saddle point

43. $3x^2 - y$, $-x + 2y$, $\frac{1}{12}$, $(0,0)$, $\left(\frac{1}{6}, \frac{1}{12}\right)$, $\left(\frac{1}{6}, \frac{1}{12}\right)$,   1,   1, $x^3 - x + 1$, $\frac{1}{\sqrt{3}}$, $\left(1 - \frac{2}{3\sqrt{3}}\right)$,   1,

1, x^3, $(0,0)$, $(1,1)$,   0, $(0,0)$

13.9   LAGRANGE MULTIPLIERS

$\boxed{\text{OBJECTIVE}}$:   Using the method of Lagrange multipliers, solve extremal
problems for functions of several independent variables
subject to one constraint equation.

44.   To determine the minimum distance from the origin to the
hyperbola $xy = 1$, we would want to minimize the function

$$f(x,y) = x^2 + y^2$$

subject to the constraint

$$g(x,y) = xy - 1 = 0.$$

We let $H(x,y,\lambda) = x^2 + y^2 - \lambda(xy - 1)$. Then, $H_x = \underline{\hspace{2cm}}$
$= 0$, $H_y = 2y - \lambda x = 0$, $H_\lambda = \underline{\hspace{2cm}} = 0$. From the first two
of these equations we obtain $x = \frac{1}{2}\lambda y$ and $y = \underline{\hspace{2cm}}$, and
subsequent substitution into the last equation $xy = 1$ gives
$\underline{\hspace{2cm}} = 1$. Since $xy = 1$ this yields $\lambda^2 = 4$ or, $\lambda = \pm 2$.
Now, $0 = xy - 1 = x(\frac{1}{2}\lambda x) - 1 = \frac{\lambda x^2}{2} - 1$ implies $x^2 = \frac{2}{\lambda}$. Thus,
λ cannot be negative for real x. Hence, $\lambda = \underline{\hspace{1cm}}$. Then
$x^2 = \frac{2}{\lambda} = 1$ or $x = \pm 1$. Since $xy = 1$ this provides the
points $(1,1)$ and $\underline{\hspace{2cm}}$. From the geometry of the
problem it is clear that there is a minimum distance from the
origin to the hypberbola, and it has the value

$$D = \sqrt{f(1,1)} = \sqrt{f(-1,-1)} = \underline{\hspace{1cm}}.$$

Notice that there is no maximum distance.

45.   Determine the radius r and the height h of the cylinder of
maximum surface area which can be inscribed in a sphere of
radius a.
Solution. We want to maximize the surface area $f(r,h) = 2\pi rh$
of the inscribed cylinder subject to the constraint

$$r^2 + \left(\frac{h}{2}\right)^2 = a^2$$

as in the figure at the right.
Assume that both $r > 0$ and
$h > 0$ so we do have an
inscribed cylinder. Let

$$H(r,h,\lambda) = 2\pi rh - \lambda\left(r^2 + \frac{h^2}{4} - a^2\right).$$

Then,

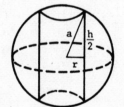

$H_r = \underline{\hspace{3cm}} = 0$, $H_h = 2\pi r - \frac{1}{2}h\lambda = 0$,   and

$H_\lambda = \underline{\hspace{3cm}} = 0$.   If $\lambda = 0$,   the equations $H_r = H_h = 0$

---

44. $2x - \lambda y$, $-xy + 1$, $\frac{1}{2}\lambda x$, $\frac{1}{4}\lambda^2 xy$,   2,   (-1,-1), $\sqrt{2}$

would require both $r = 0$ and $h = 0$, contrary to the assumption that we do have an inscribed cylinder. Thus, neither r, h, nor λ is zero. Also, from the equations $H_r = 0$ and $H_h = 0$, $\lambda r = \pi h$ and $\lambda h = 4\pi r$. Division of these last equations gives

$$\frac{\lambda r}{\lambda h} = \frac{\pi h}{\underline{\hspace{0.7cm}}} \quad \text{or,} \quad 4r^2 = \underline{\hspace{2cm}} \; .$$

Substitution into the constraint equation then yields $r^2 + \underline{\hspace{1.2cm}} = a^2$ or, $r = \underline{\hspace{1.5cm}}$ (since $r > 0$). Then, $h = 2r = \underline{\hspace{1.5cm}}$. The geometric nature of the problem ensures that the point $(r,h) = \left(\frac{a}{\sqrt{2}}, \sqrt{2}a\right)$ yields a maximum surface area

$$f\left(\frac{a}{\sqrt{2}}, \sqrt{2}a\right) = \underline{\hspace{2cm}} \; .$$

---

45. $2\pi h - 2r\lambda$, $-r^2 - \frac{h^2}{4} + a^2$, $4\pi r$, h^2, r^2, $\frac{a}{\sqrt{2}}$, $\sqrt{2}a$, $2\pi a^2$

CHAPTER 13 SELF-TEST

1. Determine, if possible, the following limits. If the limit exists, say what it is; if it does not exist, say why not.

    (a) $\lim\limits_{(x,y)\to(-1,1)} \dfrac{4x^2 - 2y + 3}{x^2 - y^2 - 1}$ (b) $\lim\limits_{(x,y)\to(0,0)} \dfrac{y^2}{x^2 + y^2}$

    (c) $\lim\limits_{(x,y)\to(0,0)} \dfrac{x - x\cos y}{2y\sin x}$

2. (a) Sketch the surface $w = f(x,y) = x^2 - 2$ in space.
    (b) Sketch the level curves of the function $f(x,y) = y - x^2 + x$ for $w = 0$, $w = 2$, and $w = -2$.
    (c) Find the largest possible xy-domain on which $w = \ln\frac{x}{y}$ is a real variable.

3. Find all first- and second-order partial derivatives.
    (a) $w = 3e^{2x}\cos y$ (b) $f(x,y,z) = zx^y$, $x > 0$

4. Find the normal line to the surface $f(x,y) = e^x \sin y$ at the point $P_0(1,\frac{\pi}{2},e)$.

5. Find an equation of the tangent plane to the hyperboloid $4x^2 - 9y^2 - 9z^2 - 36 = 0$ at $P_0(3\sqrt{3},2,2)$.

6. (a) Find the directional derivative of the function $f = z\tan^{-1}\frac{x}{y}$ at the point $P_0(1,1,2)$, and in the direction $\vec{A} = -6\vec{i} + 2\vec{j} - 3\vec{k}$.
    (b) What is the maximum value of the directional derivatives of f at P_0, and the direction of the maximum rate of change?

7. Find the indicated derivatives using the chain rule.
    (a) $\frac{dw}{dt}$, for $w = t^2 + x\sin y$, $x = \frac{1}{t}$, $y = \tan^{-1} t$
    (b) $\frac{\partial^2 w}{\partial v^2}$, for $w = \sin(4x + 5y)$, $x = u + v$, $y = u - v$

8. If $w = x + f(xy)$, where f is a differentiable function, show that $x\frac{\partial w}{\partial x} - y\frac{\partial w}{\partial y} = x$.

9. Test the surface $z = xy^2 - 2xy + 2x^2 - 15x$ for maxima, minima, and saddle points.

10. Find the absolute maximum and absolute minimum of $f(x,y) = 48xy - 32x^3 - 24y^2$ in the region bounded by the square $0 \le x \le 1$ and $0 \le y \le 1$.

11. Find the change along the plane tangent to the surface $f(x,y) = \frac{x - y}{x + y}$ at $P_0(-1, 2, -3)$, when $\Delta x = -0.1$ and $\Delta y = 0.02$. Compare this value with the increment Δf.

12. Find a linear approximation to the function $f(x,y) = xye^{x+y}$ near the point $(\frac{1}{2}, -\frac{1}{2})$.

13. Estimate the change in the volume
$$V = \frac{1}{3}\pi h(R^2 + Rr + r^2)$$
of the frustrum of a cone if the upper radius r is decreased from 4 to 3.8 cm, the base radius R is increased from 6 to 6.1 cm, and the height is increased from 8 to 8.3 cm.

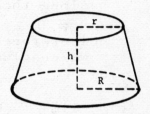

14. Suppose that the production of a certain commodity depends on two inputs, say A and B. The amounts of these are given by 100x and 100y, whose prices per unit are, respectively, \$4 and \$1. The amount of output is given by 100z, the price per unit of which is \$9. Furthermore, the production function is given by

$$f(x,y) = 5 - \frac{1}{x} - \frac{1}{y} .$$

Determine the greatest profit.

SOLUTIONS TO CHAPTER 13 SELF-TEST

1. (a) $\displaystyle\lim_{(x,y)\to(-1,1)} \frac{4x^2 - 2y + 3}{x^2 - y^2 - 1} = \frac{4(-1)^2 - 2(1) + 3}{(-1)^2 - (1)^2 - 1} = \frac{5}{-1} = -5.$

    (b)   Along the line $y = x$, $f(x,y) = f(x,x) = \dfrac{x^2}{x^2 + x^2} = \dfrac{1}{2}$,
    $x \neq 0$.

    Along the line $y = 2x$, $f(x,y) = f(x,2x) = \dfrac{4x^2}{x^2 + 4x^2} = \dfrac{4}{5}$,
    $x \neq 0$.
    Therefore, the limits along different paths do not agree so
    we conclude that f has no limit as $(x,y) \to (0,0)$.

    (c) $\displaystyle\lim_{(x,y)\to(0,0)} \frac{x - x \cos y}{2y \sin x} = \lim_{(x,y)\to(0,0)} \frac{x(1 - \cos y)}{2 \sin x \cdot y}$

    $\displaystyle = \lim_{x\to 0} \frac{x}{2 \sin x} \cdot \lim_{y\to 0} \frac{1 - \cos y}{y} = \frac{1}{2} \cdot 0 = 0.$

2. (a)   For every value of y, f is
    the parabola
    $$f(x,y) = x^2 - 2 .$$
    The graph of this cylinder is
    sketched at the right.

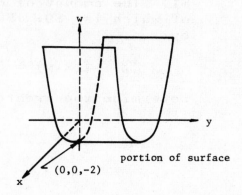

portion of surface

$(0,0,-2)$

    (b)   For any fixed $w = c$,
    $y - x^2 + x = c$ or,
    $y - c = x^2 - x$ gives
    $$y - \left(c - \frac{1}{4}\right) = \left(x - \frac{1}{2}\right)^2 ,$$

    which is a parabola.   The
    level curves $w = 0, 2, -2$
    are sketched at the right.

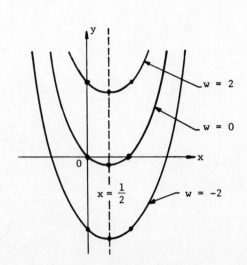

$w = 2$

$w = 0$

$x = \frac{1}{2}$

$w = -2$

(c) For $w = \ln \frac{x}{y}$ to be real-valued it is necessary that $y \neq 0$
and that $\frac{x}{y} > 0$. Therefore, the domain is the set of points
(x,y) where x and y are both positive, or both negative
(i.e., the first and third quadrants of the xy-plane
excluding the axes).

3. (a) $\frac{\partial w}{\partial x} = 6e^{2x} \cos y$, $\frac{\partial w}{\partial y} = -3e^{2x} \sin y$, $\frac{\partial^2 w}{\partial x \partial y} = \frac{\partial^2 w}{\partial y \partial x} = -6e^{2x} \sin y$,

$\frac{\partial^2 w}{\partial x^2} = 12e^{2x} \cos y$, $\frac{\partial^2 w}{\partial y^2} = -3e^{2x} \cos y$

(b) $f_x = yzx^{y-1}$, $f_y = zx^y \ln x$, $f_z = x^y$, $f_{xx} = y(y-1)zx^{y-2}$,

$f_{xy} = zx^{y-1} + yzx^{y-1} \ln x$, $f_{xz} = yx^{y-1}$, $f_{yx} = zx^{y-1} + yzx^{y-1} \ln x$,

$f_{yy} = zx^y (\ln x)^2$, $f_{yz} = x^y \ln x$, $f_{zx} = yx^{y-1}$, $f_{zy} = x^y \ln x$,

$f_{zz} = 0$

4. Notice that the point P_0 does lie on the surface since its
coordinates satisfy the equation $w = e^x \sin y$. Next,
$f_x\left(1, \frac{\pi}{2}\right) = e^x \sin y\big|_{\left(1, \frac{\pi}{2}\right)} = e$, and $f_y\left(1, \frac{\pi}{2}\right) = e^x \cos y\big|_{\left(1, \frac{\pi}{2}\right)} = 0$.

Thus, the vector $\vec{N} = e\vec{i} + 0\vec{j} - \vec{k}$ is normal to the surface at
P_0. The normal line can be represented parametrically by
$$x = 1 + et, \quad y = \frac{\pi}{2}, \quad z = e - t.$$

5. Let $g(x,y,z) = 4x^2 - 9y^2 - 9z^2$. Notice that P_0 does lie on the
given surface since $g(3\sqrt{3}, 2, 2) = 36$. Next,

$g_x(3\sqrt{3}, 2, 2) = 8x\big|_{P_0} = 24\sqrt{3}$,

$g_y(3\sqrt{3}, 2, 2) = -18y\big|_{P_0} = -36$,

$g_z(3\sqrt{3}, 2, 2) = -18z\big|_{P_0} = -36$.

The vector grad $g(P_0) = 24\sqrt{3}\,\vec{i} - 36\vec{j} - 36\vec{k}$ is normal to the
surface at P_0. Hence,
$$24\sqrt{3}(x - 3\sqrt{3}) - 36(y - 2) - 36(z - 2) = 0$$
or,
$$2\sqrt{3}x - 3y - 3z = 6$$
is an equation of the tangent plane.

6. (a) $f_x = z \cdot \dfrac{1}{1 + (x/y)^2} \cdot \dfrac{1}{y} = \dfrac{zy}{x^2 + y^2}$,

$f_y = z \cdot \dfrac{1}{1 + (x/y)^2} \cdot \left(\dfrac{-x}{y^2}\right) = \dfrac{-zx}{x^2 + y^2}$, and

$$f_z = \tan^{-1} \frac{x}{y}$$

Then, $\vec{\nabla} f(P_0) = f_x(1,1,2)\vec{i} + f_y(1,1,2)\vec{j} + f_z(1,1,2)\vec{k}$

$$= \vec{i} - \vec{j} + \frac{\pi}{4}\vec{k}.$$

A unit vector in the direction of $\vec{A}$ is

$$\vec{u} = \frac{\vec{A}}{|\vec{A}|} = \frac{\vec{A}}{\sqrt{36 + 4 + 9}} = -\frac{6}{7}\vec{i} + \frac{2}{7}\vec{j} - \frac{3}{7}\vec{k} .$$

Thus, the directional derivative of f in the direction $\vec{A}$ at P_0 is

$$\vec{\nabla} f \cdot \vec{u} = -\frac{6}{7} - \frac{2}{7} - \frac{3\pi}{28} = -\frac{32 + 3\pi}{28} \approx -1.48.$$

(b)   The direction of the maximum rate of change is in the direction of the gradient, $\vec{\nabla} f$. The value of the directional derivative in that direction is

$$|\vec{\nabla} f| = \sqrt{1 + 1 + \left(\frac{\pi}{4}\right)^2} = \frac{1}{4}\sqrt{32 + \pi^2} \approx 1.62.$$

7.   (a)   Let $z = t$ so that $w = z^2 + x \sin y$. Then,

$$\frac{dw}{dt} = \frac{\partial w}{\partial x}\frac{dx}{dt} + \frac{\partial w}{\partial y}\frac{dy}{dt} + \frac{\partial w}{\partial z}\frac{dz}{dt}$$

$$= (\sin y)\left(-t^{-2}\right) + (x \cos y)\left(\frac{1}{1 + t^2}\right) + (2z)(1)$$

$$= -t^{-2} \sin(\tan^{-1} t) + \frac{1}{t(1 + t^2)} \cos(\tan^{-1} t) + 2t.$$

(b) $\frac{\partial w}{\partial v} = \frac{\partial w}{\partial x}\frac{\partial x}{\partial v} + \frac{\partial w}{\partial y}\frac{\partial y}{\partial v} = 4 \cos(4x + 5y) - 5 \cos(4x + 5y)$

Then, setting $r = \frac{\partial w}{\partial v} = -\cos(4x + 5y)$,

$$\frac{\partial^2 w}{\partial v^2} = \frac{\partial}{\partial v}\left(\frac{\partial w}{\partial v}\right) = \frac{\partial r}{\partial v} = \frac{\partial r}{\partial x}\frac{\partial x}{\partial v} + \frac{\partial r}{\partial y}\frac{\partial y}{\partial v}$$

$$= 4 \sin(4x + 5y) - 5 \sin(4x + 5y) = -\sin(4x + 5y)$$

$$= -\sin(9u - v).$$

8.   Let $u = xy$, so that $w = x + f(u)$. Then,

$$\frac{\partial w}{\partial x} = 1 + f'(u) \frac{\partial u}{\partial x} = 1 + yf'(xy), \quad \text{and}$$

$$\frac{\partial w}{\partial y} = f'(u) \frac{\partial u}{\partial y} = xf'(xy).$$

Therefore,

$$x \frac{\partial w}{\partial x} - y \frac{\partial w}{\partial y} = x\Big(1 + yf'(xy)\Big) = y\Big(xf'(xy)\Big) = x, \quad \text{as claimed.}$$

9. Let $f(x,y) = xy^2 - 2xy + 2x^2 - 15x$. Then $f_x = y^2 - 2y + 4x - 15$
    and $f_y = 2xy - 2x$. Thus, $f_y = 2x(y - 1) = 0$, if $x = 0$ or
    $y = 1$. If $f_x = 0$ and $x = 0$, then $y^2 - 2y - 15 = 0$, or
    $(y - 5)(y + 3) = 0$. Thus, $(0,5)$ and $(0,-3)$ are critical
    points. If $f_x = 0$ and $y = 1$, then $1 - 2 + 4x - 15 = 0$, or
    $x = 4$. Thus, $(4,1)$ is a critical point. We test these
    critical points with $f_{xx} = 4$, $f_{xy} = 2y - 2$, and $f_{yy} = 2x$.

    At $(0,-3)$: $f_{xx}(0,-3) = 4$, $f_{xy}(0,-3) = -8$, and $f_{yy}(0,-3) = 0$.
    Then $f_{xx}f_{yy} - f_{xy}^2 = -64 < 0$, so the surface has a
    <u>saddle</u> point at $(0,-3)$.

    At $(0,5)$: $f_{xx}(0,5) = 4$, $f_{xy}(0,5) = 8$, and $f_{yy}(0,5) = 0$. Then
    $f_{xx}f_{yy} - f_{xy}^2 = -64 < 0$, so again the surface has a
    <u>saddle</u> point at $(0,5)$.

    At $(4,1)$: $f_{xx}(4,1) = 4$, $f_{xy}(4,1) = 0$, and $f_{yy}(4,1) = 8$. Then
    $f_{xx}f_{yy} - f_{xy}^2 = 32 > 0$ and $A > 0$, so the surface has a
    <u>local</u> <u>minimum</u> at $(4,1)$.

10. We first consider the points interior to the square:
    $$\frac{\partial f}{\partial x} = 48y - 96x^2 = 0$$
    $$\frac{\partial f}{\partial y} = 48x - 48y = 0$$
    Solving simultaneously, $y = x$ and $48x - 96x^2 = 0$ or
    $x = 0$, $x = \frac{1}{2}$.

    Thus, $(0,0)$ and $(\frac{1}{2},\frac{1}{2})$ are critical points (one of which
    occurs on the boundary of the specified region). Next,
    $$f(0,0) = 0 \quad \text{and} \quad f(\tfrac{1}{2},\tfrac{1}{2}) = \frac{48}{4} - \frac{32}{8} - \frac{24}{4} = 2 .$$
    We consider the four boundaries of the region:

    (a) $f(x,0) = -32x^3$ has a <u>minimum</u> value of -32 when $x = 1$
        and a <u>maximum</u> value of 0 when $x = 0$.

    (b) $f(1,y) = 48y - 32 - 24y^2$. Now,
        $$\frac{d}{dy}(48y - 32 - 24y^2) = 48 - 48y = 0 \quad \text{implies} \quad y = 1.$$
        When $y = 1$ we have $f(1,1) = -8$ which is a local
        <u>maximum</u> for $f(1,y)$. The <u>minimum</u> value is $f(1,0) = -32$.

    (c) $f(x,1) = 48x - 32x^3 - 24$. Here,
        $$\frac{d}{dx}(48x - 32x^3 - 24) = 48 - 96x^2 = 0 \quad \text{implies} \quad x = \pm \frac{\sqrt{2}}{2}.$$
        We must have $x \geq 0$ so we pick $x = \frac{\sqrt{2}}{2}$. Then,
        $f\left(\frac{\sqrt{2}}{2}, 1\right) = 16\sqrt{2} - 24$ is a local <u>maximum</u>. Moreover,
        $f(0,1) = -24$ is the absolute <u>minimum</u> on this boundary.

    (d) $f(0,y) = -24y^2$ which has a <u>minimum</u> value of -24 when
        $y = 1$ and a <u>maximum</u> value of 0 when $y = 0$.

    Putting all of this together yields $-32 = f(1,0)$ as the
    <u>absolute</u> <u>minimum</u> and $2 = f(\frac{1}{2},\frac{1}{2})$ as the <u>absolute</u> <u>maximum</u>.

11. $\Delta L = \frac{\partial w}{\partial x} \Delta x + \frac{\partial w}{\partial y} \Delta y = \frac{2y}{(x+y)^2} \Delta x - \frac{2x}{(x+y)^2} \Delta y$

Thus for the point $(x_0, y_0) = (-1, 2)$, $\Delta x = -0.1$ and $\Delta y = 0.02$, we find

$$df = \Delta L = \frac{4}{1}(-0.1) + \frac{2}{1}(0.02) = -0.36 \ .$$

On the other hand, the increment giving the change along the surface $w = f(x, y)$ itself is

$$\Delta f = f(-1.1, 2.02) - f(-1, 2) = \frac{-1.1 - 2.02}{-1.1 + 2.02} - \frac{-1 - 2}{-1 + 2}$$
$$= \frac{-3.12}{0.92} + \frac{3}{1} = -0.39.$$

12.   Calculating the partial derivatives

$$f_x(1, -1) = ye^{x+y} + xye^{x+y}\Big|_{\left(\frac{1}{2}, -\frac{1}{2}\right)} = -\frac{1}{2} - \frac{1}{4} = -\frac{3}{4}, \quad \text{and}$$

$$f_y(1, -1) = xe^{x+y} + xye^{x+y}\Big|_{\left(\frac{1}{2}, -\frac{1}{2}\right)} = \frac{1}{2} - \frac{1}{4} = \frac{1}{4} \ .$$

Thus, a linear approximation near $\left(\frac{1}{2}, -\frac{1}{2}\right)$ is given by

$$L(x, y) = f\left(\frac{1}{2}, -\frac{1}{2}\right) + f_x\left(\frac{1}{2}, -\frac{1}{2}\right)\left(x - \frac{1}{2}\right) + f_y\left(\frac{1}{2}, -\frac{1}{2}\right)\left(y + \frac{1}{2}\right)$$
$$= -\frac{1}{4} + \left(-\frac{3}{4}\right)\left(x - \frac{1}{2}\right) + \frac{1}{4}\left(y + \frac{1}{2}\right), \quad \text{or} \quad L(x, y) = \frac{1}{4}(1 - 3x + y).$$

13. $\Delta V \approx dV = \frac{\partial V}{\partial r} \Delta r + \frac{\partial V}{\partial R} \Delta R + \frac{\partial V}{\partial h} \Delta h$

$= \frac{1}{3}\pi h(R + 2r)\Delta r + \frac{1}{3}\pi h(2R + r)\Delta R + \frac{1}{3}\pi(R^2 + Rr + r^2)\Delta h.$
At $r = 4$, $R = 6$, $h = 8$, $\Delta r = -0.2$, $\Delta R = 0.1$, and $\Delta h = 0.3$, we find

$$\Delta V \approx \left(\frac{112\pi}{3}\right)(-0.2) + \left(\frac{128\pi}{3}\right)(0.1) + \left(\frac{76\pi}{3}\right)(0.3) = 4.4\pi.$$
The volume increases by about $4.4\pi \approx 13.8$ cubic centimeters.

14.   We wish to maximize the profit function

$$P(x, y, z) = 9(100z) - 4(100x) - 100y$$

subject to the constraint given by the equation $z = 5 - \frac{1}{x} - \frac{1}{y}$.
Thus, let

$h(x, y, z, \lambda) = 900z - 400x - 100y - \lambda\left(\frac{1}{x} + \frac{1}{y} + z - 5\right) \ .$
Setting $h_x = h_y = h_z = h_\lambda = 0$ gives the equations

$$-400 + \frac{\lambda}{x^2} = 0$$

$$-100 + \frac{\lambda}{y^2} = 0$$

$$900 - \lambda = 0$$

$$-\frac{1}{x} - \frac{1}{y} - z + 5 = 0$$

Solving simultaneously, $\lambda = 900$, $x^2 = \frac{900}{400}$ and $y^2 = \frac{900}{100}$. Thus, $x = \frac{3}{2}$, $y = 3$, and from the last equation, $z = 4$ (all variables must be positive here). Therefore,

$$P\left(\tfrac{3}{2}, 3, 4\right) = 900 \cdot 4 - 400 \cdot \tfrac{3}{2} - 100 \cdot 3 = 2700.$$

Because x and y are both in the interval $(0, \infty)$, and P is a negative number when x and y are either close to zero or very large, we conclude that this corresponds to a <u>maximum profit</u> of \$2700.

NOTES:

CHAPTER 14 MULTIPLE INTEGRALS

14.1 DOUBLE INTEGRALS

OBJECTIVE A: Evaluate a given (double) iterated integral, and sketch the region over which the integration extends.

1. $\int_0^1 \int_{x^3}^{x^2} xy \, dy \, dx$

$= \int_0^1 \left[\underline{\hspace{2cm}} \right]_{y=x^3}^{y=x^2} dx$

$= \int_0^1 \left(\frac{1}{2}x^5 - \underline{\hspace{2cm}} \right) dx$

$= \frac{1}{12}x^6 - \underline{\hspace{2cm}} \right]_{x=0}^{x=1}$

$= \frac{1}{12} - \underline{\hspace{2cm}} = \underline{\hspace{2cm}}.$

Sketch the region of integration in the coordinate system provided.

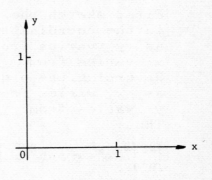

2. $\int_0^\pi \int_0^{y^2} \sin \frac{x}{y} \, dx \, dy$

$= \int_0^\pi \left[\underline{\hspace{2cm}} \right]_{x=0}^{x=y^2} dy$

$= \int_0^\pi \left(-y \cos y + \underline{\hspace{1cm}} \right) dy$

$= \left(-y \sin y \right]_{y=0}^{y=\pi} + \int_0^\pi \sin y \, dy \right)$

$\qquad + \underline{\hspace{2cm}} \right]_{y=0}^{y=\pi}$

$= 0 + \left(\underline{\hspace{1cm}} \right) \right]_{y=0}^{y=\pi} + \frac{\pi^2}{2} = \underline{\hspace{2cm}} \approx 6.93.$

Sketch the region of integration.

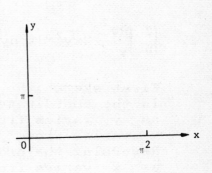

---

1. $\frac{1}{2}xy^2$, $\frac{1}{2}x^7$,

$\frac{1}{16}x^8$, $\frac{1}{16}$,

$\frac{1}{48}$

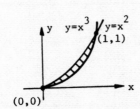

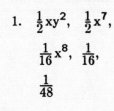

2. $-y \cos \frac{x}{y}$, y,

$\frac{1}{2}y^2$, $-\cos y$,

$\frac{\pi^2}{2} + 2$

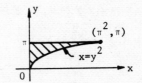

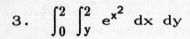 : Given a (double) iterated integral, write an equivalent
double iterated integral with the order of integration
reversed. Sketch the region over which the integration
takes place and evaluate the new integral.

3. $\int_0^2 \int_y^2 e^{x^2} \, dx \, dy$

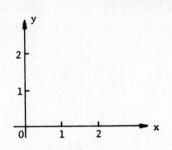

First sketch the region of integration
in the coordinate system at the right:
as y varies from y = 0 to y = 2,
x varies from _____.
Reversing the order of integration:
as x varies from x = 0 to x = 2,
y varies from _____.
Thus,

$$\int_0^2 \int_y^2 e^{x^2} \, dx \, dy = \int_{\underline{}}^{\underline{}} \int_{\underline{}}^{\underline{}} e^{x^2} \, dy \, dx$$

$$= \int_0^2 \underline{}]_{y=0}^{y=x} \, dx = \int_0^2 \underline{} \, dx$$

$$= \underline{}]_{x=0}^{x=2} = \underline{} \approx 26.80 .$$

4. $\int_0^2 \int_y^{6-y^2} 2xy^2 \, dx \, dy$

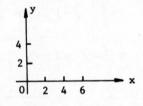

First sketch the region of integration
in the coordinate system at the right:
as y varies from y = 0 to y = 2,
x varies from _____.
Reversing the order of integration,
as x varies from x = 0 to x = 2,
y varies from _____;
but as x varies from x = 2 to
x = 6, y varies from _____.
Thus, the iterated integral becomes the sum of two iterated
integrals when the order of integration is reversed:

---

3. x = y to x = 2, y = 0 to y = x,

$\int_0^2 \int_0^x ,$ $ye^{x^2},$ $xe^{x^2},$ $\frac{1}{2}e^{x^2},$

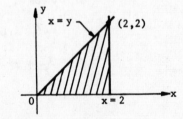

$\frac{1}{2}\left(e^4 - 1\right)$

$$\int_0^2 \int_y^{6-y^2} 2xy^2 \ dx \ dy = \int \underline{\quad} \int \underline{\quad} 2xy^2 \ dy \ dx + \int \underline{\quad} \int \underline{\quad} 2xy^2 \ dy \ dx$$

$$= \int_0^2 \underline{\qquad\qquad}]_{y=0}^{y=x} \ dx + \int_2^6 \tfrac{2}{3}xy^3]\underline{\quad} \ dx$$

$$= \int_0^2 \tfrac{2}{3}x^4 \ dx + \int_2^6 \tfrac{2}{3}x(6-x)^{3/2} \ dx$$

$$= \underline{\qquad\qquad}]_{x=0}^{x=2} + \int_4^0 -\tfrac{2}{3}(6-u)u^{3/2} \ du$$
$$(\text{where} \quad u = 6 - x)$$

$$= \underline{\qquad\quad} - \tfrac{2}{3}\left(\tfrac{12}{5}u^{5/2} - \tfrac{2}{7}u^{7/2}\right)]_{u=4}^{u=0} \approx 31.09.$$

OBJECTIVE C :   Find the volume of a solid whose base is a specified region A in the xy-plane, and whose top is a given surface $z = f(x,y)$

5.   The base of a solid is the region in the xy-plane that is bounded by the ellipse $x^2 + 9y^2 = 36$ for $y \le 0$. Sketch this region in the coordinate system at the right. The top of the solid is the plane $z = -y$. Thus, the volume of the solid is computed as the following double integral:

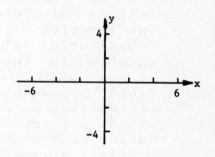

$$V = \int \underline{\quad} \int \underline{\quad} -y \ dy \ dx = \int_{-6}^6 \underline{\qquad\qquad}]_{y=\underline{\quad}}^{y=\underline{\quad}} \ dx$$

$$= \tfrac{1}{2} \int_{-6}^6 \underline{\qquad\qquad} \ dx = \tfrac{1}{18}(\underline{\qquad\qquad})]_{x=-6}^{x=6} = \underline{\qquad\qquad}.$$

---

4. $x = y$ to $x = 6 - y^2$, $y = 0$ to $y = x$,
   $y = 0$ to $y = \sqrt{6 - x}$,

   $\int_0^2 \int_0^x + \int_2^6 \int_0^{\sqrt{6-x}}, \quad \tfrac{2}{3}xy^3,$

   $]_{y=0}^{y=\sqrt{6-x}}, \quad \tfrac{2}{15}x^5, \quad \tfrac{64}{15}$

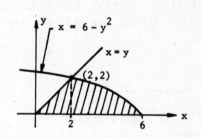

5. $\int_{-6}^6 \int_{-\frac{1}{3}\sqrt{36-x^2}}^0,$

   $-\tfrac{1}{2}y^2]_{y=-\frac{1}{3}\sqrt{36-x^2}}^{y=0},$

   $\tfrac{1}{9}(36 - x^2), \quad 36x - \tfrac{1}{3}x^3, \quad 16$

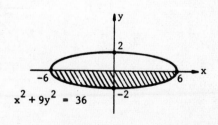

$x^2 + 9y^2 = 36$

14.2   AREAS, MOMENTS, AND CENTERS OF MASS

OBJECTIVE A : For a specified region   R   in the xy-plane
      (a)   sketch the region   R;
      (b)   label each bounding curve of the region with its equation, and find the coordinates of the boundary points where the curves intersect; and
      (c)   find the area of the region by evaluating an appropriate (double) iterated integral.

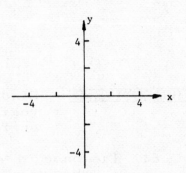

6.   Let us find the area of the region   R   in the xy-plane bounded by the circle $x^2 + y^2 = 16$ and the parabola $y^2 = 6x$.   First sketch the region   R   and label the bounding curves in the coordinate system at the right.   Notice that we seek the area <u>inside</u> both curves because   x   must be positive in order that $y^2 = 6x$ hold for real numbers.   Next we calculate the points of intersection of the circle and parabola.   Substituting $y^2 = 6x$ into the equation for the circle gives

   _____ = 16   or, $(x + 8)($_____$) = 0$.   Thus,   x = _____ since   x   cannot be negative.   Then $y^2 = 6x = $ _____.   This yields the two points   _____   and   _____   as the boundary points where the two curves intersect.   We now want to decide what order of integration we should choose.   We see that <u>vertical</u> strips sometimes go from the lower branch of the parabola to the upper branch (if $0 \leq x \leq 2$),   but sometimes from the lower semi-circle to the upper semi-circle (if $2 \leq x \leq 4$).   Thus, integration in the order of first   y   and then   x   requires that the area be taken in two separate pieces.

   On the other hand, <u>horizontal</u> strips always go from the parabola as the left boundary over to the circle on the right, and the area is given by the integral

$$A = \int \underline{\quad\quad} \int \underline{\quad\quad\quad} \, dx \, dy \ .$$

---

6. $x^2 + 6x$, $x - 2$, 2, 12, $(2, -2\sqrt{3})$

$(2, 2\sqrt{3})$, $\displaystyle\int_{-2\sqrt{3}}^{2\sqrt{3}}$, $\displaystyle\int_{y^2/6}^{\sqrt{16 - y^2}}$

$\sqrt{16 - y^2} - \dfrac{1}{6} y^2$

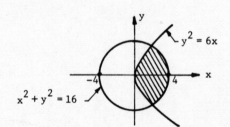

Evaluation of this integral gives the result

$$A = \int_{-2\sqrt{3}}^{2\sqrt{3}} \left(\underline{\hspace{3cm}} \right) \, dy$$

$$= \frac{1}{2}\left(y\sqrt{16 - y^2} + 16 \sin^{-1} \frac{y}{4}\right) - \frac{1}{18} y^3 \Big]_{y=-2\sqrt{3}}^{y=2\sqrt{3}}$$

$$= \frac{4\sqrt{3}}{3} + 8 \sin^{-1} \frac{\sqrt{3}}{2} - 8 \sin^{-1}\left(-\frac{\sqrt{3}}{2}\right)$$

$$= \frac{4\sqrt{3}}{3} + 8\left(\frac{\pi}{3}\right) - 8\left(-\frac{\pi}{3}\right) = \frac{4\sqrt{3} + 16\pi}{3} \approx 19.06.$$

7.   Find the area of the region in the first quadrant that is bounded by the curves $x + y = 4$ and $y = \frac{3}{x}$.

<u>Solution</u>.   Sketch the region in the coordinate system at the right, and label the bounding curves.   To calculate the points of intersection of the line and the hyperbola, substitute $y = 4 - x$ into $xy = 3$ which gives the result $\underline{\hspace{2cm}}$, or $(x - 3)(\underline{\hspace{1cm}}) = 0$.   This yields the two points $(3,1)$ and $\underline{\hspace{1.5cm}}$ as the boundary points where the two curves intersect.   In this example, either order of integration leads to a single double integral giving the area.   If we use <u>horizontal</u> strips, the area is given by

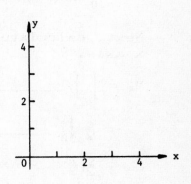

$$A = \int_{\underline{\hspace{0.5cm}}}^{\overline{\hspace{0.5cm}}} \int_{\underline{\hspace{0.5cm}}}^{\overline{\hspace{0.5cm}}} \, dx \, dy \; .$$

On the other hand, using <u>vertical</u> strips the area is given by

$$A = \int_{\underline{\hspace{0.5cm}}}^{\overline{\hspace{0.5cm}}} \int_{\underline{\hspace{0.5cm}}}^{\overline{\hspace{0.5cm}}} \, dy \, dx \; .$$

Evaluation of either integral (we take the latter) gives

$$A = \int_{1}^{3} \left(\underline{\hspace{3cm}} \right) \, dx$$

$$= 4x - \frac{1}{2}x^2 - 3 \ln x \Big]_{x=1}^{x=3} = 4 - 3 \ln 3 \approx 0.704.$$

---

7. $x(4 - x) = 3,$ $x - 1,$ $(1,3),$

$$\int_{1}^{3} \int_{3/y}^{4-y} dx \, dy, \quad \int_{1}^{3} \int_{3/x}^{4-x} dy \, dx,$$

$4 - x - \frac{3}{x}$

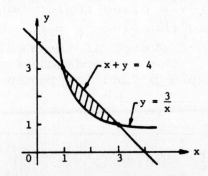

OBJECTIVE B : Given a plane region   R   in the xy-plane, find its center of mass.

8.   Consider the region   R   in the xy-plane bounded by the curves $y^2 = x$ and $y = x$. The region   R   is sketched in the figure at the right. To find the <u>center of mass</u> we assume the region has constant uniform density $\delta(x,y) \equiv c$. First, calculate the mass of the region. This is given by the double integral,

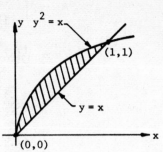

$$M = \int\int \delta(x,y) \; dA = \int_{\underline{\quad}}^{\overline{\quad}} \int_{\underline{\quad}}^{\overline{\quad}} c \; dy \; dx$$

$$= \int_0^1 \underline{\hspace{3cm}} dx = c\left(\tfrac{2}{3}x^{3/2} - \tfrac{1}{2}x^2\right)\Big]_{x=0}^{x=1} = \underline{\hspace{3cm}}.$$

Next, calculate the moment of the mass with respect to the x-axis,

$$M_x = \int\int \underline{\hspace{3cm}} = \int_0^1 \int_x^{\sqrt{x}} cy \; dy \; dx$$

$$= \tfrac{c}{2} \int_0^1 \underline{\hspace{2cm}}\Big]_{y=x}^{y=\sqrt{x}} dx = \tfrac{c}{2} \int_0^1 (x - x^2) \; dx$$

$$= \tfrac{c}{2}\left(\tfrac{1}{2}x^2 - \tfrac{1}{3}x^3\right)\Big]_{x=0}^{x=1} = \underline{\hspace{3cm}}.$$

The moment of the mass with respect to the y-axis is,

$$M_y = \int\int x\delta(x,y) \, dA = \int_{\underline{\quad}}^{\overline{\quad}} \int_{\underline{\quad}}^{\overline{\quad}} cx \; dy \; dx$$

$$= c \int_0^1 x(\sqrt{x} - x) \; dx = c(\underline{\hspace{3cm}})\Big]_{x=0}^{x=1} = \underline{\hspace{3cm}}.$$

We then obtain the coordinates of the center of mass,

$$\bar{x} = \frac{M_y}{M} = \underline{\hspace{2cm}} \quad \text{and} \quad \bar{y} = \frac{M_x}{M} = \underline{\hspace{2cm}}.$$

OBJECTIVE C : Given a plane region   R   in the xy-plane, find its moments of inertia about various axes.

9.   Find the moments of inertia about the x-axis and about the y-axis for the plane region bounded by the parabola $y = kx^2$, $k > 0$,   and the line $y = b$, $b > 0$.
<u>Solution</u>. A sketch of the region is given in the figure on the next page. The x-coordinates of the points of intersection of the line $y = b$ and the parabola $y = kx^2$ are $x = -\sqrt{b/k}$ and $x = \underline{\hspace{2cm}}$.

---

8. $\int_0^1 \int_x^{\sqrt{x}}$, $c(\sqrt{x} - x)$, $\tfrac{c}{6}$, $y\,\delta(x,y) \, dA$, y^2, $\tfrac{c}{12}$, $\int_0^1 \int_x^{\sqrt{x}}$, $\tfrac{2}{5}x^{5/2} - \tfrac{1}{3}x^3$, $\tfrac{c}{15}$, $\tfrac{2}{5}$, $\tfrac{1}{2}$

We assume that the region is of constant uniform density $\delta(x,y) \equiv 1$. Then the moment of inertia about the x-axis is given by

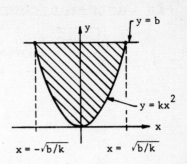

$$I_x = \iint y^2 \delta(x,y) \; dA$$

$$= \int \underline{\hspace{1cm}} \int \underline{\hspace{1cm}} y^2 \; dx \; dy$$

$$= \frac{2}{\sqrt{k}} \int_0^b \underline{\hspace{2cm}} \; dy = \underline{\hspace{2cm}} \; .$$

The moment of inertia about the y-axis is

$$I_y = \int \int \underline{\hspace{3cm}} = \int_0^b \int_{-\sqrt{y/k}}^{\sqrt{y/k}} x^2 \; dx \; dy$$

$$= \frac{2}{3k^{3/2}} \int_0^b \underline{\hspace{2cm}} \; dy = \underline{\hspace{2cm}} \; .$$

If we calculate the mass of the plane region,

$$M = \int \int \delta(x,y) \; dA = \int_0^b \int_{-\sqrt{y/k}}^{\sqrt{y/k}} dx \; dy = 2 \int_0^b \sqrt{y/k} \; dy = \frac{4b^{3/2}}{4\sqrt{k}} \; ,$$

then we can express the moments of inertia in terms of the mass as follows:

$$I_x = \frac{3b^2}{7} M \quad \text{and} \quad I_y = \underline{\hspace{2cm}} \; .$$

14.3 DOUBLE INTEGRALS IN POLAR FORM

OBJECTIVE A: Given a (double) iterated integral in Cartesian coordinates, change it to an equivalent double integral in polar coordinates and then evaluate the integral thus obtained.

10. $\displaystyle\int_0^{a/\sqrt{2}} \int_y^{\sqrt{a^2-y^2}} \sin(x^2 + y^2) \; dx \; dy$

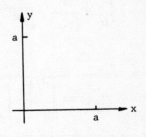

First sketch the region of integration in the coordinate system at the right. In converting the integral to polar coordinates we need to find the polar limits of integration for the region.
We see that θ varies from $\theta = \underline{\hspace{1.5cm}}$

to $\theta = \frac{\pi}{4}$ radians, and r varies from $r = 0$ to $r = \underline{\hspace{1.5cm}}$ units.
Thus, the integral becomes

---

9. $\sqrt{b/k}$, $\displaystyle\int_0^b \int_{-\sqrt{y/k}}^{\sqrt{y/k}}$, $y^{5/2}$, $\dfrac{4b^{7/2}}{7\sqrt{k}}$, $x^2\delta(x,y) \; dA$, $y^{3/2}$, $\dfrac{4b^{5/2}}{15k^{3/2}}$, $\dfrac{b}{5k} M$

<u>In Cartesian coordinates</u>   <u>In polar coordinates</u>

$$\int_0^{a/\sqrt{2}} \int_y^{\sqrt{a^2-y^2}} \sin(x^2 + y^2)dx\,dy \;=\; \int_0^{\pi/4} \int_0^{\underline{}} \underline{}\, dr\,d\theta$$

$$=\; \int_0^{\pi/4} \underline{}\Big]_{r=0}^{r=a}\, d\theta$$

$$=\; \int_0^{\pi/4} \left(\tfrac{1}{2} - \tfrac{1}{2}\cos a^2\right)\, d\theta$$

$$=\; \underline{}\,.$$

11. $\displaystyle \int_0^1 \int_{\sqrt{1-x^2}}^{\sqrt{2-x^2}} \frac{1}{\sqrt{x^2 + y^2}}\, dy\,dx$

A sketch of the region of integration
is shown at the right. Notice that
the line $y = x$ intersects the
circle $x^2 + y^2 = 2$ at the point
$(1,1)$. Let us find the polar
limits of integration for this region.
The region can be described as the
union of two regions: as θ varies

from $\theta = 0$ to $\theta = \frac{\pi}{4}$, r varies
from $r = $ \underline{} (on the circle
$x^2 + y^2 = 1$) to the line $x = 1$ or
$r \sec \theta$ (since $x = r \cos \theta = 1$);
as θ varies from $\theta = \frac{\pi}{4}$ to $\theta = \frac{\pi}{2}$, r varies from

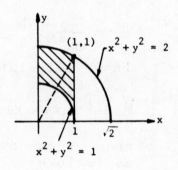

\underline{} (from the circle $x^2 + y^2 = 1$ to the circle
$x^2 + y^2 = 2$). Therefore, the integral becomes

<u>In Cartesian coordinates</u>

$$\int_0^1 \int_{\sqrt{1-x^2}}^{\sqrt{2-x^2}} \frac{1}{\sqrt{x^2 + y^2}}\, dy\,dx$$

<u>In polar coordinates</u>

$$=\; \int_0^{\pi/4} \int_{\underline{}}^{\underline{}} \left(\tfrac{1}{r}\right) r\,dr\,d\theta + \int_{\underline{}}^{\underline{}} \int_{\underline{}}^{\underline{}} \left(\tfrac{1}{r}\right)\,dr\,d\theta$$

---

10. 0, a, $\displaystyle\int_0^{\pi/4} \int_0^a \sin(r^2)\, r\,dr\,d\theta,$

$-\tfrac{1}{2}\cos(r^2),\quad \tfrac{\pi}{4}\left(\tfrac{1}{2} - \tfrac{1}{2}\cos a^2\right)$

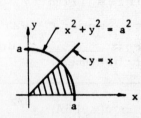

$$= \int_0^{\pi/4} (\underline{\hspace{2cm}}) \; d\theta + \int_{\pi/4}^{\pi/2} (\sqrt{2} - 1) \; d\theta$$

$$= \ln |\sec \theta + \tan \theta| - \theta]_{\theta=0}^{\theta=\pi/4} + \underline{\hspace{3cm}}]_{\theta=\pi/4}^{\theta=\pi/2}$$

$$= \left(\ln |\sec \tfrac{\pi}{4} + \tan \tfrac{\pi}{4}| - \tfrac{\pi}{4} - 0\right) + \underline{\hspace{3cm}}$$

$$= \ln\left(\sqrt{2} + 1\right) + \tfrac{\pi}{4}\left(\sqrt{2} - 2\right) \approx 0.42.$$

OBJECTIVE B :   In an applied problem involving double integration (e.g., finding an area, volume, center of mass or moment of inertia), express the double integral in polar coordinates (when appropriate, to make the integrations easier), and then evaluate the integral thus obtained.

12.   Let us find the area of one petal of the rose $r = a \sin 3\theta$ sketched in the figure at the right.   The petal in the first quadrant is traced out as θ varies from $\theta = 0$ to $\theta = \frac{\pi}{3}$:   r starts at $r = 0$,   reaches its maximum value $r = a$ when $\theta = \underline{\hspace{1.5cm}}$,   and returns to $r = 0$ when $\theta = \frac{\pi}{3}$.   Thus, as θ varies from $\theta = 0$ to $\theta = \frac{\pi}{3}$,   r varies from $\underline{\hspace{2.5cm}}$.   The area of the petal is given in terms of polar coordinates by

$$A = \int \int \; dA = \int_{\underline{\hspace{0.7cm}}} \int_{\underline{\hspace{0.7cm}}} r \; dr \; d\theta$$

$$= \int_0^{\pi/3} \underline{\hspace{2cm}}]_{r=0}^{r=a\sin\theta} \; d\theta = \int_0^{\pi/3} \underline{\hspace{3cm}} \; d\theta$$

$$= \tfrac{a^2}{4} \int_0^{\pi/3} (1 - \cos 6\theta) \; d\theta = \tfrac{a^2}{4} [\underline{\hspace{3cm}}]_{\theta=0}^{\theta=\pi/3} = \underline{\hspace{2cm}} \; .$$

---

11.   1, $r = 1$ to $r = \sqrt{2}$, $\int_0^{\pi/4} \int_1^{\sec \theta} + \int_{\pi/4}^{\pi/2} \int_1^{\sqrt{2}}$, $\sec \theta - 1$, $(\sqrt{2} - 1)\theta$, $\tfrac{\pi}{4}(\sqrt{2} - 1)$

12. $\tfrac{\pi}{6}$, $r = 0$ to $r = a \sin 3\theta$ (the boundary of the petal), $\int_0^{\pi/3} \int_0^{a\sin 3\theta}$, $\tfrac{1}{2} r^2$, $\tfrac{1}{2} a^2 \sin^2 3\theta$, $\theta - \tfrac{1}{6} \sin 6\theta$, $\tfrac{\pi a^2}{12}$

13.   Let us find the center of mass of the semi-circular region bounded by the x-axis and the curve $y = \sqrt{a^2 - x^2}$, when the density of the region is proportional to the square of its distance from the center of the semi-circular arc: $\delta(x,y) = k\left(x^2 + y^2\right)$ for some proportionality constant k. The region is sketched in the figure at the right. In terms of polar coordinates, the mass of the region is given by

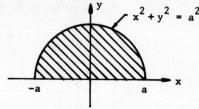

$$M = \int\int \delta(x,y)\ dA = \int_{\rule{1.5em}{0.4pt}} \int_{\rule{1.5em}{0.4pt}} kr^2 \cdot r\ dr\ d\theta$$

$$= \int_0^\pi \frac{k}{4} r^4 \Big]_{r=0}^{r=a}\ d\theta = \underline{\hspace{4cm}}\ .$$

The moment of the mass with respect to the x-axis in terms of polar coordinates is

$$M_x = \int\int y\delta(x,y)\ dA = \int_0^\pi \int_0^a \underline{\hspace{3cm}} \cdot r\ dr\ d\theta$$

$$= \int_0^\pi \underline{\hspace{3cm}}\ d\theta = \frac{2ka^5}{5}\ .$$

From the symmetry of the plate and its density function across the y-axis, it is clear that M_y is zero. (This may also be verified by direct calculation.) Thus,

$$\bar{x} = \underline{\hspace{2cm}} \quad \text{and} \quad \bar{y} = \underline{\hspace{2.5cm}}$$

give the coordinates for the center of mass.

14.   Suppose we wish to find the volume of the ellipsoid $9x^2 + 9y^2 + z^2 = 9$. In Cartesian coordinates the volume is given by

$$V = 2\int_{-1}^1 \int_{-\sqrt{1-x^2}}^{\sqrt{1-x^2}} z\ dy\ dz, \quad \text{where} \quad z = 3\sqrt{1 - x^2 - y^2}$$

(the factor 2 occurs because we selected $z \geq 0$, and the ellipsoid is symmetric across the plane $z = 0$). We see that this iterated integral would be very cumbersome to calculate. Let us try the integration in polar coordinates: $x = r\cos\theta$ and $y = r\sin\theta$ transform the equation $9x^2 + 9y^2 + z^2 = 9$ into the polar equation $\underline{\hspace{3cm}}$. Thus, when $z = 0$ (in the xy-plane), $r = \underline{\hspace{1.5cm}}$. Therefore, in terms of polar coordinates the volume is given by

---

13. $\int_0^\pi \int_0^a,\ \frac{\pi ka^4}{4},\ kr^2(r\sin\theta),\ \frac{ka^5}{5}\sin\theta,\ 0,\ \frac{8a}{5\pi}$

$$V = 2 \int_{\underline{\quad}} \int_{\underline{\quad}} 3\sqrt{1 - r^2} \cdot r \, dr \, d\theta$$

$$= 2 \int_0^{2\pi} \underline{\hspace{3cm}} \Big]_{r=0}^{r=1} d\theta = 2 \int_0^{2\pi} \underline{\quad} d\theta = 4\pi.$$

14.4 TRIPLE INTEGRALS IN RECTANGULAR COORDINATES. VOLUMES AND AVERAGE VALUES

OBJECTIVE A : Evaluate an iterated triple integral.

15. $\displaystyle\int_0^2 \int_0^x \int_0^{1-2x+y} e^z \, dz \, dy \, dx$

$$= \int_0^2 \int_0^x e^z \Big|_{z=0}^{z=1-2x+y} dy \, dx$$

$$= \int_0^2 \int_0^x \underline{\hspace{5cm}} dy \, dx$$

$$= \int_0^2 \left[e^{1-2x+y} - \underline{\hspace{2cm}} \right]_{y=0}^{y=x} dx$$

$$= \int_0^2 \left[\left(\underline{\hspace{3cm}} \right) - e^{1-2x} \right] dx$$

$$= -e^{1-x} - \tfrac{1}{2}x^2 + \underline{\hspace{2cm}} \Big]_{x=0}^{x=2}$$

$$= \left(e^{-1} - 2 + \tfrac{1}{2}e^{-3} \right) - \left(\underline{\hspace{2cm}} \right)$$

$$\approx -0.984 \; .$$

OBJECTIVE B : By triple integration, find the volume of a specified region D in xyz-space.

16. Let us calculate the volume of the region in space bounded by the cylinders $z = x^2$ and $z = 4 - x^2$, and the planes $y = 0$, $y = 5$. A sketch of the region is shown at the right.
The two surfaces intersect when $4 - x^2 = x^2$, or $x^2 = 2$; that is they intersect on the plane

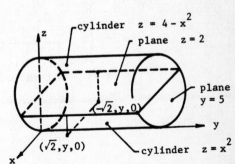

cylinder $z = 4 - x^2$
plane $z = 2$
plane $y = 5$
$(-\sqrt{2}, y, 0)$
$(\sqrt{2}, y, 0)$
cylinder $z = x^2$

---

14. $z^2 + 9r^2 = 9$, 1, $\displaystyle\int_0^{2\pi} \int_0^1$, $-\left(1 - r^2\right)^{3/2}$

15. $e^{1-2x+y} - 1$, y, $e^{1-x} - x$, $\tfrac{1}{2}e^{1-2x}$, $-e + \tfrac{1}{2}e$

$z = 2$. The volume projects into the region R in the xy-plane described by $-\sqrt{2} \leq x \leq \sqrt{2}$ and $0 \leq y \leq 5$. Thus, the volume is given by the triple iterated integral,

$$V = \int_{\underline{}}^{\overline{}} \int_{\underline{}}^{\overline{}} \int_{\underline{}}^{\overline{}} dz\, dy\, dx$$

$$= \int_{-\sqrt{2}}^{\sqrt{2}} \int_{0}^{5} \underline{\hspace{3cm}} dy\, dx = \int_{-\sqrt{2}}^{\sqrt{2}} \underline{\hspace{3cm}}\Big]_{y=0}^{y=5} dx$$

$$= \int_{-\sqrt{2}}^{\sqrt{2}} \underline{\hspace{3cm}} dx = \underline{\hspace{3cm}}\Big]_{x=-\sqrt{2}}^{x=\sqrt{2}} = \underline{\hspace{2cm}} .$$

14.5   MASSES AND MOMENTS IN THREE DIMENSIONS

OBJECTIVE: By triple integration, find the mass, the center of mass, or the moments of inertia of an object distributed over a region D of xyz-space and having density $\delta = \delta(x,y,z)$ at the point (x,y,z) of D.

17.   Find the center of mass of the homogeneous solid bounded below by the cone $x^2 + y^2 = z^2$ and above by the plane $z = 4$.

Solution. The solid region is sketched at the right. Since the region is geometrical we take the density to be unity, $\delta \equiv 1$. By symmetry $\bar{x} = \underline{\hspace{1.5cm}}$ and $\bar{y} = \underline{\hspace{1.5cm}}$. To find $\bar{z}$ we first calculate

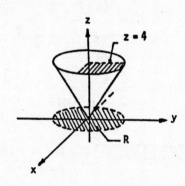

$$M_{xy} = \iiint\limits_{R} \int_{z=0}^{z=4} z\delta\, dz\, dy\, dx$$

$$= \iint\limits_{R} \left[\tfrac{1}{2}z^2\right]_{z=0}^{z=4} dy\, dx$$

$$= \iint\limits_{R} \underline{\hspace{2.5cm}} dy\, dx$$

$$= 8 \cdot \text{area of planar region } R$$

$$= \underline{\hspace{2.5cm}} .$$

16. $\displaystyle\int_{-\sqrt{2}}^{\sqrt{2}} \int_{0}^{5} \int_{x^2}^{4-x^2}$, $4 - 2x^2$, $y(4 - 2x^2)$, $5(4 - 2x^2)$, $20x - \tfrac{10}{3}x^3$, $\dfrac{80\sqrt{2}}{3}$

The mass of the object is

$$M = \iiint_R \int_{z=0}^{z=4} \delta \; dz \; dy \; dx$$

$$= \iint_R \underline{\hspace{2cm}} \; dy \; dx$$

$$= \underline{\hspace{2cm}} \; .$$

Therefore,

$$\overline{z} = \frac{M_{xy}}{M} = \underline{\hspace{2cm}} \; .$$

The center of mass is $\underline{\hspace{4cm}}$.

14.6   TRIPLE INTEGRALS IN CYLINDRICAL AND SPHERICAL COORDINATES

OBJECTIVE A : Find by triple integration, in cylindrical coordinates, the volume of a specified region in space. (The region may be described in Cartesian coordinates.)

18.   Consider the region in space that lies inside the sphere $x^2 + y^2 + z^2 = 9$, but <u>outside</u> the cylinder $x^2 + y^2 = 1$. The region is sketched in the figure below. The z-axis is an axis of symmetry so that cylindrical coordinates should be useful for finding the volume of the region by triple integration. The equation of the sphere in cylindrical coordinates is $z^2 = \underline{\hspace{2cm}}$. The projection of the volume within the sphere and outside the cylinder in the xy-plane is a washer with inner radius $r = 1$ and outer radius $r = \underline{\hspace{1cm}}$. This is the shaded shown in the diagram, and in polar coordinates it is described by the inequalities

$\underline{\hspace{1cm}} \leq r \leq \underline{\hspace{1cm}}$ and
$\underline{\hspace{1cm}} \leq \theta \leq \underline{\hspace{1cm}}$. For a
point (x,y), or (r,θ),
in this washer, the altitude
z of the solid region varies
from $z = -\sqrt{9 - r^2}$ (the
surface of the sphere below
the xy-plane) to $z = \underline{\hspace{2cm}}$
(that portion of the surface
above the xy-plane). Then the volume of the solid region in cylindrical coordinates is given by

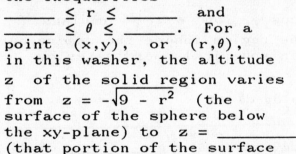

$$V = \int_1^3 \int_0^{2\pi} \int_{-\sqrt{9-r^2}}^{\underline{\qquad}} \underline{\hspace{2cm}} = \int_1^3 \int_0^{2\pi} \underline{\hspace{2cm}} \, d\theta \; dr$$

$$= \int_1^3 4\pi r \sqrt{9 - r^2} \; dr = \underline{\hspace{2cm}} \Big]_{r=1}^{r=3} = \frac{64\pi\sqrt{2}}{3} \approx 94.78.$$

| OBJECTIVE B |: Find by triple integration, in spherical coordinates, the volume of a specified region in space. (The region may be described in Cartesian coordinates.)

19. Consider the region in space that lies inside the sphere $x^2 + y^2 + z^2 = 4$, but outside the cone $x^2 + y^2 = z^2$. The region is sketched in the figure at the right. The origin is a point of symmetry of the solid so that spherical coordinates should be useful for finding the volume of the region by triple integration.

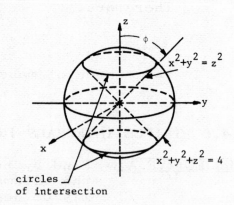

circles
of intersection

To find the intersection of the sphere and the cone, we substitute $x^2 + y^2 = z^2$ into the equation of the sphere obtaining $z^2 + z^2 = 4$ or $z^2 = \underline{\hspace{1cm}}$. Thus, $z = -\sqrt{2}$, or $z = \sqrt{2}$. The plane $z = -\sqrt{2}$ gives the circle of intersection $x^2 + y^2 = 2$ on the sphere which lies $\sqrt{2}$ units <u>below</u> the xy-plane; and $z = \sqrt{2}$ provides the corresponding circle of intersection above the xy-plane. In spherical coordinates, $z = \rho \cos \phi$, so the plane $z = \sqrt{2}$ corresponds to $\sqrt{2} = \rho \cos \phi$. Since the circles of intersection lie on the sphere where $\rho = 2$, the angle ϕ (see figure) for the upper circle satisfies the equation $\sqrt{2} = 2 \cos \phi$. Thus, $\phi = \underline{\hspace{1cm}}$ radians. Similarly, the angle for the lower circle satisfies $-\sqrt{2} = 2 \cos \phi$, or $\phi = \frac{3\pi}{4}$ radians. Therefore, the volume of the region inside the sphere but outside the cone is given in spherical coordinates by

$$V = \int_{\underline{\quad}}^{\underline{\quad}} \int_{\underline{\quad}}^{\underline{\quad}} \int_{\underline{\quad}}^{\underline{\quad}} \rho^2 \sin \phi \; d\rho \; d\theta \; d\phi$$

$$= \int_{\pi/4}^{3\pi/4} \int_0^{2\pi} \frac{1}{3}\rho^3 \sin \phi \Big]_{\rho=0}^{\rho=2} d\theta \; d\phi = \int_{\pi/4}^{3\pi/4} \int_0^{2\pi} \underline{\hspace{2cm}} d\theta \; d\phi$$

$$= \int_{\pi/4}^{3\pi/4} \underline{\hspace{2cm}} d\phi = \frac{16\pi\sqrt{2}}{3} \approx 23.70.$$

---

18. $9 - r^2$, 3, $1 \le r \le 3$, $0 \le \theta \le 2\pi$, $\sqrt{9 - r^2}$, $\int_1^3 \int_0^{2\pi} \int_{-\sqrt{9-r^2}}^{\sqrt{9-r^2}} dz \; r \; d\theta \; dr$, $2r\sqrt{9 - r^2}$, $-\frac{4\pi}{3}\big(9 - r^2\big)^{3/2}$

19. 2, $\frac{\pi}{4}$, $\int_{\pi/4}^{3\pi/4} \int_0^{2\pi} \int_0^2$, $\frac{8}{3} \sin \phi$, $\frac{16\pi}{3} \sin \phi$

OBJECTIVE C : By triple integration, using cylindrical or spherical coordinates, find the mass, the center of mass, or the moments of inertia of an object distributed over a region D of xyz-space and having density $\delta = \delta(x,y,z)$ at the point (x,y,z) of D.

20. Let us find the center of mass of a solid hemisphere of radius a if the density at any point is proportional to the distance of the point from the axis of the solid. We choose coordinates so that the origin is at the center of the sphere, the z-axis is the axis of the hemisphere, and consider the hemisphere that lies above the xy-plane. In spherical coordinates the hemisphere may be described by $\rho = a$, $0 \leq \phi \leq \frac{\pi}{2}$, and _____ $\leq \theta \leq$ _____. The distance $r = \rho \sin \phi$ is measured from the z-axis parallel to the xy-plane, and we are given that $\delta = k\rho \sin \phi$ for some constant k of proportionality. Therefore, the mass of the solid is given by the triple integral (in spherical coordinates),

$$M = \int \underline{\quad} \int \underline{\quad} \int \underline{\quad} (k\rho \sin \phi) \; \rho^2 \sin \phi \; d\rho \; d\phi \; d\theta$$

$$= \int_0^{2\pi} \int_0^{\pi/2} \underline{\hspace{3cm}} d\phi \; d\theta$$

$$= \frac{ka^4}{4} \int_0^{2\pi} \left[\frac{\phi}{2} - \frac{1}{4} \sin 2\phi\right]_{\phi=0}^{\phi=\pi/2} d\theta = \underline{\hspace{3cm}} .$$

Because of the symmetry of the solid about the z-axis in both shape and density, $\bar{x} = \bar{y} = 0$. We calculate $\bar{z}$:

$$\iiint z\delta \; dV = \int_0^{2\pi} \int_0^{\pi/2} \int_0^a \underline{\hspace{5cm}} d\rho \; d\phi \; d\theta$$

$$= \int_0^{2\pi} \int_0^{\pi/2} \underline{\hspace{4cm}} d\phi \; d\theta$$

$$= \int_0^{2\pi} \frac{ka^5}{15} \sin^3\phi \Big]_{\phi=0}^{\phi=\pi/2} d\theta = \underline{\hspace{2cm}} .$$

Therefore, $\bar{z} = \frac{2\pi ka^5}{15M} = \underline{\hspace{2cm}} .$

---

20. $0 \leq \theta \leq 2\pi$, $\int_0^{2\pi} \int_0^{\pi/2} \int_0^a$, $\frac{1}{4}ka^4 \sin^2\phi$, $\frac{ka^4\pi^2}{8}$, $(\rho \cos \phi)(k\rho \sin \phi)\rho^2 \sin \phi$, $\frac{1}{5} ka^5 \sin^2\phi \cos \phi$, $\frac{2\pi ka^5}{15}$, $\frac{16a}{15\pi}$

21.   Find the center of mass of the homogenous solid inside the paraboloid $x^2 + y^2 = z$ and <u>outside</u> the cone $x^2 + y^2 = z^2$.

<u>Solution</u>. The solid region is sketched at the right. A slice through the solid by a plane parallel to the xy-plane produces a washer-shaped area like that shaded area shown in the figure. The cone and the paraboloid intersect when $z = z^2$, or $z = 0$ (the base) or $z =$ _____ (the top rim of the solid). Since the region is geometrical we take the density as unity. Because the z-axis is an axis of symmetry it will be convenient to use cylindri-cal coordinates. In cylindrical

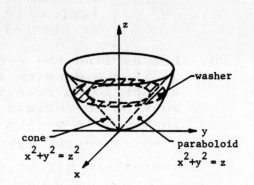

coordinates, the equation $x^2 + y^2 = z^2$ of the cone becomes _____ and the equation for the paraboloid becomes _____. The projection of the solid into the xy-plane is the interior of a circle centered at the origin of radius $r =$ _____ (since $r = z$ for the top rim of the solid). In polar coordinates this circular disk is expressed by the inequalities _____ $\leq r \leq$ _____ and _____ $\leq \theta \leq$ _____. For a point (x,y), or (r,θ), in this disk the altitude of the solid region varies from the paraboloid $z =$ _____ to the surface of the cone $z =$ _____, both expressed in terms of cylindrical coordinates. Therefore, the mass of the solid is given by the triple integral

$$M = \iiint dV = \int \underline{\quad} \int \underline{\quad} \int \underline{\quad} dz\ r\ dr\ d\theta$$

$$= \int_0^{2\pi} \int_0^1 \underline{\qquad\qquad}\ dr\ d\theta = \int_0^{2\pi} \underline{\qquad\qquad}]_{r=0}^{r=1}\ d\theta = \frac{\pi}{6}\ .$$

From the symmetry of the solid, it is clear that the center of mass must lie on the z-axis. Thus, $\bar{x} =$ _____ and $\bar{y} =$ _____. We calculate $\bar{z}$.

$$\iiint z\ dV = \int \underline{\quad} \int \underline{\quad} \int \underline{\quad} z\ dz\ r\ dr\ d\theta$$

$$= \int_0^{2\pi} \int_0^1 \underline{\qquad}]_{z=r^2}^{z=r}\ dr\ d\theta = \frac{1}{2} \int_0^{2\pi} \int_0^1 (r^3 - r^5)\ dr\ d\theta$$

$$= \frac{1}{2} \int_0^{2\pi} \left(\frac{1}{4}r^4 - \frac{1}{6}r^6\right)]_{r=0}^{r=1}\ d\theta = \frac{\pi}{12}.$$

Therefore, $\bar{z} =$ _____ .

---

21. 1, $r^2 = z^2$, $r^2 = z$, 1, $0 \leq r \leq 1$, $0 \leq \theta \leq 2\pi$, r^2, r, $\int_0^{2\pi} \int_0^1 \int_{r^2}^{r}$, $r(r - r^2)$,

$\frac{1}{3}r^3 - \frac{1}{4}r^4$, 0, 0, $\int_0^{2\pi} \int_0^1 \int_{r^2}^{r}$, $\frac{1}{2}rz^2$, $\frac{1}{2}$

22. To calculate the moment of inertia about the z-axis of the solid in Problem 21 above, we find

$$I_z = \iiint \left(x^2 + y^2\right)\delta \; dV \qquad (\text{where} \quad \delta \equiv 1)$$

$$= \int \underline{\quad} \int \underline{\quad} \int \underline{\quad} \; \underline{\hspace{2cm}} \; dz \; r \; dr \; d\theta$$

$$= \int_0^{2\pi} \int_0^1 \underline{\hspace{2.5cm}} \; dr \; d\theta$$

$$= \int_0^{2\pi} \underline{\hspace{3cm}}]_{r=0}^{r=1} \; d\theta = \underline{\hspace{2cm}} \; .$$

14.7 SUBSTITUTIONS IN MULTIPLE INTEGRALS

23. If $f(x,y)$, $x = g(u,v)$, and $y = h(u,v)$ are continuous and have continuous partial derivatives, then substitution for x and y in the double integral

$$\iint_R f(x,y) \; dx \; dy$$

results in the integral

$$\iint_G f(g(u,v), h(u,v)) |J(u,v)| \; du \; dv$$

where

$$J(u,v) = \frac{\partial(x,y)}{\partial(u,v)} = \underline{\hspace{3cm}} \; .$$

Here the region G in the uv-plane is transformed $\underline{\hspace{2cm}}$ into the region R in the xy-plane by the functions g and $\underline{\hspace{1cm}}$.

---

22. $\int_0^{2\pi} \int_0^1 \int_{r^2}^r \; r^2, \quad r^3(r - r^2), \quad \frac{1}{5}r^5 - \frac{1}{6}r^6, \quad \frac{\pi}{15}$

23. $\begin{vmatrix} \dfrac{\partial x}{\partial u} & \dfrac{\partial x}{\partial v} \\[2mm] \dfrac{\partial y}{\partial u} & \dfrac{\partial y}{\partial v} \end{vmatrix}$, one-to-one, h

OBJECTIVE: Evaluate a double integral over a specified region R in the xy-plane by changing variables $x = g(u,v)$ and $y = h(u,v)$ and integrating over a region G in the uv-plane.

24.   Let R be the region in the first quadrant of the xy-plane bounded by the parallel lines $y = 2x - 2$, $y = 2x + 1$, and the lines $x + y = 1$, $x + y = 3$. A sketch of R is shown in the figure at the right. Use the transformation $u = x + y$, $v = y - 2x$ with $u \geq 1$ and $v \geq -2$ to evaluate

$$\iint_R \frac{3x}{x + y} \, dy \, dx .$$

Solution. First solve the transformation equations for x and y in terms of u and v. Routine algebra gives

$$x = \underline{\hspace{2cm}} , \quad y = \underline{\hspace{2cm}} .$$

The Jacobian of the transformation is

$$J(u,v) = \begin{vmatrix} \underline{\hspace{0.5cm}} & \underline{\hspace{0.5cm}} \\ \underline{\hspace{0.5cm}} & \underline{\hspace{0.5cm}} \end{vmatrix} = \frac{1}{3} .$$

The boundaries of the region R correspond to: $x + y = u = 1$, $x + y = u = \underline{\hspace{1cm}}$, $y - 2x = v = -2$, $y - 2x = v = \underline{\hspace{1cm}}$. Thus substitution in the double integral gives

$$\iint_R \frac{3x}{x + y} \, dy \, dx = \int_1^3 \int_{-2}^1 \underline{\hspace{2cm}} \cdot \frac{1}{3} \, dv \, du$$

$$= \frac{1}{3} \int_1^3 \int_{-2}^1 \left(1 - \frac{v}{u}\right) dv \, du$$

$$= \frac{1}{3} \int_1^3 \left[\underline{\hspace{1.5cm}}\right]_{v=-2}^{v=1} du$$

$$= \frac{1}{3} \int_1^3 \left(1 - \frac{1}{2u} + \underline{\hspace{1cm}}\right) du$$

$$= \int_1^3 \left(\underline{\hspace{1.5cm}}\right) du$$

$$= u + \underline{\hspace{1.5cm}}\Big]_{u=1}^{u=3}$$

$$= 3 + \frac{1}{2} \ln 3 - 1 - \frac{1}{2} \ln 1 \approx 2.55 .$$

---

24. $\frac{1}{3}(u - v)$, $\frac{1}{3}(2u + v)$, $\begin{vmatrix} \frac{1}{3} & -\frac{1}{3} \\ \frac{2}{3} & \frac{1}{3} \end{vmatrix}$, 3, 1, $\frac{u - v}{u}$, $v - \frac{1}{2u}v^2$, $2 + \frac{2}{u}$, $1 + \frac{1}{2u}$, $\frac{1}{2} \ln u$

CHAPTER 14   SELF-TEST

1.   Evaluate the following double integrals, and sketch the region over which the integration extends.

(a) $\int_1^3 \int_{-y}^{2y} xe^{y^3}\, dx\, dy$ (b) $\int_0^{\pi/2} \int_0^{\cos x} e^y \sin x\, dy\, dx$

2.   Write an equivalent double integral with the order of integration reversed for each of the following. Sketch the region over which the integration extends and evaluate the new integral.

(a) $\int_1^2 \int_{x^2}^{x^3} dy\, dx$ (b) $\int_0^2 \int_0^{\sqrt{4-y^2}} \sqrt{4 - x^2}\, dx\, dy$

3.   Find, by double integration, the volume of the solid under the paraboloid $z = x^2 + y^2$ and lying above the region in the xy-plane bounded by the curves $y = x$ and $y = x^2$.

4.   Find, by double integration, the area of the region in the xy-plane bounded by the curves $y = x^2$ and $y = 4x - x^2$.

5.   Find the center of mass of the region in the first quadrant of the xy-plane bounded by the curve $y = \sqrt{x}$ and the line $x = 4$, if the density at (x,y) is $\delta(x,y) = xy$.

6.   Find the moments of inertia about the x- and y-axes for the region in Problem 5 (with the same density).

7.   Change the double integral $\int_0^{2a} \int_0^{\sqrt{2ay-y^2}} x\, dx\, dy$ to an equivalent double integral in polar coordinates, and then evaluate.

8.   By double integration, find the volume of the solid under the hemisphere $x^2 + y^2 + z^2 = 4$, $z \geq 0$, and lying above the region in the xy-plane within the cylinder $(x - 1)^2 + y^2 = 1$.

9.   By triple integration, find the volume of the solid within the sphere $x^2 + y^2 + z^2 = z$ and above the cone $z^2 = x^2 + y^2$.

10.   Find the center of mass of the wedge cut out of the cylinder $x^2 + y^2 = 1$ by the plane $z = y$ above, and the plane $z = 0$ below.

11.   Find the moment of inertia of a right circular cone of base radius a, altitude h, and constant density $\delta = 1$, about its axis (i.e., the axis through the vertex and perpendicular to the base).

12.   Evaluate the integral

$$\iint\limits_{R} \cos\left(\frac{x - y}{x + y}\right) \, dx \, dy$$

over the region $R = \{(x,y): \ 0 \leq x \leq 1, \ 0 \leq y \leq 1, \ x + y \leq 1\}$
by means of the transformation $x - y = u$ and $x + y = v$.

SOLUTIONS TO CHAPTER 14 SELF-TEST

1. (a) The region of integration is shown in the figure at the right. Evaluation of the integral leads to,

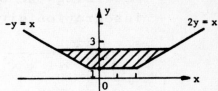

$$\int_1^3 \int_{-y}^{2y} xe^{y^3} \, dx \, dy$$

$$= \int_1^3 \tfrac{1}{2}x^2 e^{y^3} \Big]_{x=-y}^{x=2y} \, dy = \int_1^3 \tfrac{1}{2}(4y^2 - y^2)e^{y^3} dy$$

$$= \int_1^3 \tfrac{3}{2}y^2 e^{y^3} \, dy = \tfrac{1}{2}e^{y^3} \Big]_{y=1}^{y=3} = \tfrac{1}{2}\left(e^{27} - e\right) .$$

(b) The region of integration is sketched at the right. Evaluation of the integral gives

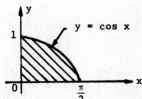

$$\int_0^{\pi/2} \int_0^{\cos x} e^y \sin x \, dy \, dx$$

$$= \int_0^{\pi/2} e^y \sin x \Big]_{y=0}^{y=\cos x} \, dx$$

$$= \int_0^{\pi/2} (\sin x \, e^{\cos x} - \sin x) \, dx = -e^{\cos x} + \cos x \Big]_{x=0}^{x=\pi/2}$$

$$= e - 2 \approx 0.718 .$$

2. (a) The region of integration is sketched at the right. As y varies from $y = 1$ to $y = 4$, x varies from $x = y^{1/3}$ to $x = y^{1/2}$; but as y varies from $y = 4$ to $y = 8$, x varies from $x = y^{1/3}$ to $x = 2$. Thus, the integral becomes the sum of two integrals when the order of integration is reversed:

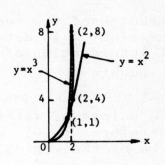

$$\int_1^2 \int_{x^2}^{x^3} dy \, dx = \int_1^4 \int_{y^{1/3}}^{y^{1/2}} dx \, dy + \int_4^8 \int_{y^{1/3}}^2 dx \, dy$$

$$= \int_1^4 (y^{1/2} - y^{1/3}) \, dy + \int_4^8 (2 - y^{1/3}) \, dy = \tfrac{17}{12} .$$

(b)   The region of integration is sketched at the right. Reversing the order of integration gives

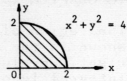

$$\int_0^2 \int_0^{\sqrt{4-y^2}} \sqrt{4 - x^2} \; dx \; dy$$

$$= \int_0^2 \int_0^{\sqrt{4-x^2}} \sqrt{4 - x^2} \; dy \; dx = \int_0^2 y\left(4 - x^2\right)^{1/2} \Big]_{y=0}^{y=\sqrt{4-x^2}} dx$$

$$= \int_0^2 (4 - x^2) \; dx = 4x - \tfrac{1}{3}x^3 \Big]_{x=0}^{x=2}$$

$$= \frac{16}{3} \; .$$

3.   The region of integration is sketched at the right. The volume is given by,

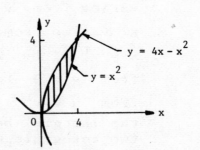

$$V = \iint z \; dA = \int_0^1 \int_{x^2}^{x} (x^2 + y^2) \; dy \; dx$$

$$= \int_0^1 (x^2 y + \tfrac{1}{3}y^3)\big]_{y=x^2}^{y=x} dx$$

$$= \int_0^1 (\tfrac{4}{3}x^3 - x^4 - \tfrac{1}{3}x^6) \; dx = \tfrac{1}{3}x^4 - \tfrac{1}{5}x^5 - \tfrac{1}{21}x^7\big]_{x=0}^{x=1}$$

$$= \frac{9}{105} \approx 0.086 .$$

4.   A sketch of the region is shown at the right. To determine the points of intersection of the two curves, we equate $y = x^2$ and $y = 4x - x^2$:

$$x^2 = 4x - x^2 \quad \text{or,} \quad 2x(x - 2) = 0 .$$

The points of intersection are therefore $(0,0)$ and $(2,4)$. The area of the region is given by

$$A = \iint dA = \int_0^2 \int_{x^2}^{4x-x^2} dy \; dx$$

$$= \int_0^2 (4x - x^2 - x^2) \; dx = \tfrac{8}{3}.$$

5. The mass of the region is given by,

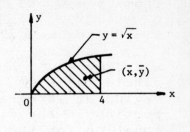

$$M = \iint \delta \ dA = \int_0^4 \int_0^{\sqrt{x}} xy \ dy \ dx$$

$$= \int_0^4 \tfrac{1}{2}xy^2 \big]_{y=0}^{y=\sqrt{x}} \ dx$$

$$= \tfrac{1}{2} \int_0^4 x^2 \ dx = \tfrac{32}{3} \ .$$

Calculation of the first moments gives,

$$M_x = \iint y\delta \ dA = \int_0^4 \int_0^{\sqrt{x}} xy^2 \ dy \ dx = \int_0^4 \tfrac{1}{3}xy^3 \big]_{y=0}^{y=\sqrt{x}} \ dx$$

$$= \tfrac{1}{3} \int_0^4 x^{5/2} \ dx = \tfrac{256}{21} \ ,$$

and

$$M_y = \iint x\delta \ dA = \int_0^4 \int_0^{\sqrt{x}} x^2y \ dy \ dx = \int_0^4 \tfrac{1}{2}x^2y^2 \big]_{y=0}^{y=\sqrt{x}} \ dx$$

$$= \tfrac{1}{2} \int_0^4 x^3 \ dx = \tfrac{256}{8} = 32 \ .$$

Therefore,

$$\bar{x} = \frac{M_y}{M} = 3 \quad \text{and} \quad \bar{y} = \frac{M_x}{M} = \frac{8}{7}.$$

6. The moment of inertia about the x-axis is given by,

$$I_x = \iint y^2\delta \ dA = \int_0^4 \int_0^{\sqrt{x}} xy^3 \ dy \ dx$$

$$= \int_0^4 \tfrac{1}{4}y^4x \big]_{y=0}^{y=\sqrt{x}} \ dx = \tfrac{1}{4} \int_0^4 x^3 \ dx = 16 \ .$$

The moment of inertia about the y-axis is given by,

$$I_y = \iint x^2\delta \ dA = \int_0^4 \int_0^{\sqrt{x}} x^3y \ dy \ dx$$

$$= \int_0^4 \tfrac{1}{2}x^3y^2 \big]_{y=0}^{y=\sqrt{x}} \ dx = \tfrac{1}{2} \int_0^4 x^4 \ dx = 102.4 \ .$$

Thus, the polar moment of inertia is given by

$$I_0 = I_x + I_y = 118.4 \ .$$

7. For the region of integration, as y varies from $y = 0$ to $y = 2a$, x varies from

    $x = 0$ to $x = \sqrt{2ay - y^2}$. The equation

    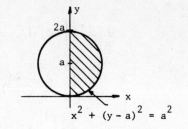

    $x = \sqrt{2ay - y^2}$ describes the right semi-circle $x^2 + (y - a)^2 = a^2$, $x \geq 0$. The region of integration is the shaded portion in the figure at the right. A polar equation for the circle is $r = 2a \sin \theta$. Then the region of integration may be described in terms of polar coordinates as follows: θ varies from $\theta = 0$ to $\theta = \frac{\pi}{2}$, and r varies from $r = 0$ to $r = 2a \sin \theta$. Thus, the double integral becomes,

    $$\int_0^{2a} \int_0^{\sqrt{2ay-y^2}} x \; dx \; dy = \int_0^{\pi/2} \int_0^{2a \sin \theta} r \cos \theta \cdot r \; dr \; d\theta$$

    $$= \int_0^{\pi/2} \frac{8a^3}{3} \sin^3 \theta \cos \theta \; d\theta = \frac{2a^3}{3} \sin^4 \theta \Big]_{\theta=0}^{\theta=\pi/2} = \frac{2a^3}{3} \; .$$

8. The Cartesian equation $z = \sqrt{4 - x^2 - y^2}$ can be written as $z = \sqrt{4 - r^2}$ using polar coordinates for x and y. The region of integration in the xy-plane is the interior of the circle $(x - 1)^2 + y^2 = 1$. In terms of polar coordinates this region is described by the inequalities $-\frac{\pi}{2} \leq \theta \leq \frac{\pi}{2}$ and $0 \leq r \leq 2 \cos \theta$. Because of the symmetry of the solid we need only compute the volume lying above the interior of the semi-circle in the first quadrant of the xy-plane, and then multiply by 2. Then the total volume is given by

    $$V = \iint z \; dA = 2 \int_0^{\pi/2} \int_0^{2 \cos \theta} \sqrt{4 - r^2} \cdot r \; dr \; d\theta$$

    $$= 2 \int_0^{\pi/2} -\frac{1}{3} \left(4 - r^2\right)^{3/2} \Big]_{r=0}^{r=2 \cos \theta} d\theta = -\frac{16}{3} \int_0^{\pi/2} (\sin^3 \theta - 1) \; d\theta$$

    $$= -\frac{16}{3} \left[-\frac{1}{3} \cos \theta (\sin^2 \theta + 2) - \theta \right]_{\theta=0}^{\theta=\pi/2} = \frac{8\pi}{3} - \frac{32}{9} \approx 4.82 \; .$$

9. The solid region resembles an ice cream cone, and is sketched below. The sphere is centered at $(0, 0, \frac{1}{2})$ and has radius $a = \frac{1}{2}$. The cone and the sphere intersect when $z^2 + z^2 = z$, or $z = 0$ (the bottom tip of the cone), or $z = \frac{1}{2}$ (at the rim of the cone). The equation of the cone can be written in spherical

coordinates as $\rho^2 \cos^2\phi = \rho^2 \sin^2\phi$ or, for $\rho \neq 0$, $\phi = \frac{\pi}{4}$ (since $0 \leq \phi \leq \pi$ by definition). In spherical coordinates the equation of the sphere is $\rho^2 = \rho \cos\phi$, or $\rho = \cos\phi$ (since $\rho \neq 0$ for the top of the region). Therefore, the volume of the region is given in spherical coordinates by the triple integral

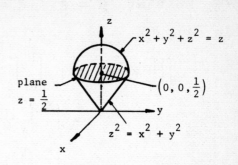

$$V = \iiint dV = \int_0^{2\pi} \int_0^{\pi/4} \int_0^{\cos\phi} \rho^2 \sin\phi \, d\rho \, d\phi \, d\theta$$

$$= \int_0^{2\pi} \int_0^{\pi/4} \tfrac{1}{3} \cos^3\phi \, \sin\phi \, d\phi \, d\theta$$

$$= -\frac{1}{12} \int_0^{2\pi} \cos^4\phi]_{\phi=0}^{\phi=\pi/4} \, d\theta = \frac{\pi}{8} \approx 0.39 \, .$$

10. A sketch of the solid is shown in the figure at the right. We will use cylindrical coordinates. The plane $z = y$ becomes $z = r \sin\theta$ in cylindrical coordinates. The region of integration in the xy-plane is the interior of the semi-circle $y = \sqrt{1 - x^2}$. In cylindrical coordinates this region is given by $0 \leq \theta \leq \pi$ and $0 \leq r \leq 1$. Since the region is geometrical we take the density as unity. Then the mass is given by,

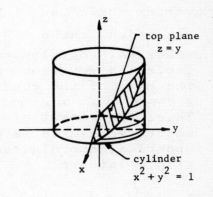

$$M = \iiint \delta \, dV$$

$$= \int_0^\pi \int_0^1 \int_0^{r\sin\theta} dz \, r \, dr \, d\theta$$

$$= \int_0^\pi \int_0^1 r^2 \sin\theta \, dr \, d\theta = \tfrac{1}{3} \int_0^\pi \sin\theta \, d\theta = \tfrac{2}{3} \, .$$

From the symmetry of the solid across the yz-plane, $\bar{x} = 0$. To find $\bar{y}$ and $\bar{z}$ we calculate the first moments:

$$\iiint y\delta \, dV = \int_0^\pi \int_0^1 \int_0^{r\sin\theta} r \sin\theta \, dz \, r \, dr \, d\theta$$

$$= \int_0^\pi \int_0^1 r^3 \sin^2\theta \, dr \, d\theta = \tfrac{1}{4} \int_0^\pi \sin^2\theta \, d\theta$$

$$= \tfrac{1}{4}\left(\tfrac{\theta}{2} - \tfrac{1}{4} \sin 2\theta\right)]_{\theta=0}^{\theta=\pi} = \frac{\pi}{8}$$

so that $\bar{y} = \frac{\pi}{8} \div \frac{2}{3} = \frac{3\pi}{16}$.

Also,

$$\iiint z\delta \, dV = \int_0^\pi \int_0^1 \int_0^{r\sin\theta} z \, dz \, r \, dr \, d\theta$$

$$= \int_0^\pi \int_0^1 \frac{1}{2} z^2 \Big]_{z=0}^{z=r\sin\theta} r \, dr \, d\theta$$

$$= \int_0^\pi \int_0^1 \frac{1}{2} r^3 \sin^2\theta \, dr \, d\theta = \frac{1}{8} \int_0^\pi \sin^2\theta \, d\theta = \frac{\pi}{16}$$

from which it follows that $\bar{z} = \frac{\pi}{16} \div \frac{2}{3} = \frac{3\pi}{32}$.

11.   We may choose the origin at the vertex of the cone with the z-axis as the axis of the cone. Then an equation for the cone in Cartesian coordinates is given by $z = \sqrt{x^2 + y^2}$, $0 \le z \le h$. A portion of the cone is shown at the right. From the geometry of the figure we see that the vertex angle α of the cone satisfies $\tan \alpha = \frac{a}{h}$. In spherical coordinates, the Cartesian equation $z = h$ for the base of the cone becomes $\rho \cos \phi = h$ or, $\rho = h \sec \phi$. Thus, the moment of inertia about the z-axis is given in spherical coordinates by the triple integral,

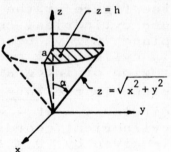

$$I_z = \iiint (x^2 + y^2) \, \delta \, dV \qquad (\delta \equiv 1)$$

$$= \int_0^{2\pi} \int_0^\alpha \int_0^{h\sec\phi} (\rho^2 \sin^2\phi) \, \rho^2 \sin\phi \, d\rho \, d\phi \, d\theta$$

$$= \frac{h^5}{5} \int_0^{2\pi} \int_0^\alpha \tan^3\phi \, \sec^2 {}_2\phi \, d\phi \, d\theta$$

$$= \frac{h^5}{5} \int_0^{2\pi} \frac{1}{4} \tan^4\alpha \, d\theta = \frac{\pi h^5}{10} \tan^4\alpha = \frac{\pi a^4 h}{10}$$

since $\alpha = \tan^{-1}\left(\frac{a}{h}\right)$.

12. The region R is sketched in
     the figure at the right.
     From the transformation
     equations

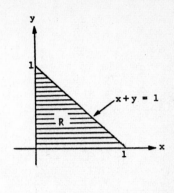

$$u = x - y \quad \text{and} \quad v = x + y$$

we find by routine algebra

$$x = \tfrac{1}{2}(u + v)$$

$$y = \tfrac{1}{2}(v - u) \ .$$

The Jacobian of the transformation is

$$J(u,v) = \frac{\partial(x,y)}{\partial(u,v)} = \begin{vmatrix} 1/2 & 1/2 \\ -1/2 & 1/2 \end{vmatrix} = \tfrac{1}{2} \ .$$

The region R is transformed to the region G in the uv-plane
as follows:

  left boundary $x = 0,\ 0 \le y \le 1$ becomes $u = -y$ and $v = y$
           or $v = -u$;

  right boundary (hypotenuse) $x + y = 1$ becomes $v = 1$;

  base boundary $y = 0,\ 0 \le x \le 1$ becomes $u = x$ and $v = x$
           or $v = u$.

The region is sketched in the
figure at the right. Thus
substitution in the integral gives

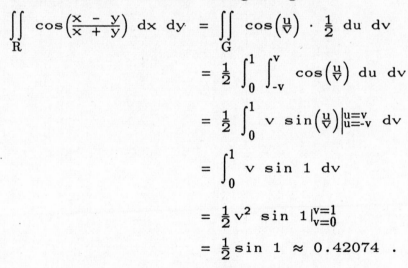

$$\iint\limits_{R} \cos\left(\frac{x - y}{x + y}\right)\ dx\ dy = \iint\limits_{G} \cos\left(\frac{u}{v}\right) \cdot \tfrac{1}{2}\ du\ dv$$

$$= \tfrac{1}{2} \int_{0}^{1} \int_{-v}^{v} \cos\left(\tfrac{u}{v}\right)\ du\ dv$$

$$= \tfrac{1}{2} \int_{0}^{1} v\ \sin\left(\tfrac{u}{v}\right)\Big|_{u=-v}^{u=v}\ dv$$

$$= \int_{0}^{1} v\ \sin 1\ dv$$

$$= \tfrac{1}{2} v^2\ \sin 1 \Big|_{v=0}^{v=1}$$

$$= \tfrac{1}{2} \sin 1 \approx 0.42074\ .$$

NOTES:

CHAPTER 15 VECTOR ANALYSIS

15.1 LINE INTEGRALS

[OBJECTIVE]: For a curve C specified by the position vector
$$\vec{r}(t) = x(t)\vec{i} + y(t)\vec{j} + z(t)\vec{k}, \quad a \le t \le b$$

evaluate the line integral
$$\int_C f(x,y,z)\ ds$$
for a given function f along C.

1. Evaluate the integral of $f(x,y,z) = x - z$ along the curve $\vec{r}(t) = t\vec{i} + (1 - t)\vec{j} + \vec{k}, \quad 0 \le t \le 1$.

   Solution. First we express ds in terms of dt:

$$ds = \sqrt{\left(\frac{dx}{dt}\right)^2 + \left(\frac{dy}{dt}\right)^2 + \left(\frac{dz}{dt}\right)^2}\ dt$$

$$= \sqrt{(1)^2 + (\underline{-1})^2 + (\underline{0})^2}\ dt$$

$$= \sqrt{2}\ dt\ .$$

   Thus,

$$\int_C f(x,y,z)\ ds = \int_{\underline{0}}^{\underline{1}} f[x(t),y(t),z(t)]\sqrt{2}\ dt$$

$$= \sqrt{2} \int_0^1 [x(t) - z(t)]dt \qquad t - 1 + t$$

$$= \sqrt{2} \cdot \underline{\int_0^1 (t-1)dt}$$

$$= \sqrt{2} \cdot \tfrac{1}{2}(t - 1)^2\big|_0^1 = \underline{-\dfrac{\sqrt{2}}{2}}\ .$$

2. Integrate $f(x,y,z) = x + y - z$ over the path from $(0,0,0)$ to $(1,1,1)$ shown in the figure on the next page.

   Solution. The path in the xy-plane is described by

$$\vec{r}_1(t) = \underline{t\vec{i} + t\vec{j}}\ , \quad 0 \le t \le 1$$

$$|\vec{v}_1(t)| = \underline{\sqrt{2}}\ .$$

---

1. $-1, \quad 0, \quad \int_0^1, \quad \int_0^1 (t - 1)\ dt, \quad -\dfrac{\sqrt{2}}{2}$

The second path is given by

$$\vec{r}_2(t) = \vec{i} + \vec{j} + t\vec{k}, \quad \underline{0} \leq t \leq \underline{1}$$

$$|\vec{v}_2(t)| = \underline{1} \;.$$

Having decided on the parameterization, we integrate $f(x,y,z) = x + y - z$ along each path in the direction of increasing t:

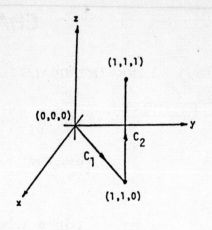

(0,0,0)   (1,1,1)   (1,1,0) C_1 C_2

$$\int_{C_1 \cup C_2} f(x,y,z)\ ds = \int_0^1 (t + t - 0)\sqrt{2}\ dt$$

$$+ \int_{\underline{0}}^{\underline{1}} \underline{(1 + 1 - t)}\ dt$$

$$= \underline{\sqrt{2}\, t^2}\Big]_0^1 + [2t - \tfrac{t^2}{2}]_0^1$$

$$= \underline{\sqrt{2}} + \tfrac{3}{2} \;.$$

15.2   VECTOR FIELDS, WORK, CIRCULATION, AND FLUX

OBJECTIVE A :   Define the term <u>vector</u> <u>field</u> and give several examples of vector fields.

3.   Let D be a region of 3-space. If, to each point P in D, a ___vector___ $\vec{F}(P)$ is assigned, the collection of all such ___vectors___ is called a vector ___field___ .

4.   For a fluid flowing with steady-state flow in a region D, a vector field is given by the assignment of a ___velocity___ vector $\vec{v}$ indicating the flow at each point of D. This vector field may be described by writing a vector equation for the velocity at each point $P(x,y,z)$ of D.

5.   If the gradient vector of a scalar function is attached to each point of a level curve, or level surface, of the function, a vector field is obtained. If the function is $z = f(x,y)$, this vector field lies in ___x-y plane___ ; if $w = f(x,y,z)$ the vector field lies ___space___ . In the latter case we have a three-dimensional vector field on ___level surface___ .

---

2. $t\vec{i} + t\vec{j}$, $\sqrt{2}$, $0 \leq t \leq 1$,   1, $\int_0^1 (1 + 1 - t) \cdot 1\ dt$, $\sqrt{2}t^2$, $\sqrt{2}$

3.   vector,   vectors,   field

4.   velocity

5.   the xy-plane,   in space,   the level surface

OBJECTIVE B :   For a specified curve C evaluate the line integral

$$\int_C \vec{F} \cdot d\vec{r}$$

where $\vec{F} = M(x,y,z)\vec{i} + N(x,y,z)\vec{j} + P(x,y,z)\vec{k}$ is a given vector field with continuous components through-out some connected region D that contains C. Assume $\vec{r} = x\vec{i} + y\vec{j} + z\vec{k}$ is the vector from the origin to the point (x,y,z).

6. If $\vec{F}$ is interpreted as a force whose point of application moves along the curve C from a point A to a point B, then the line integral

$$\int_C \vec{F} \cdot d\vec{r}$$

is the ___work___ done by the ___force___ along the curve.

7. If we calculate the dot product of the vectors $\vec{F}$ and $d\vec{r}$, an alternate form for the line integral in the Objecitve B is

$$\int_C M\,dx + N\,dy + P\,dz$$

8. Let us calculate $\int_C \vec{F} \cdot d\vec{r}$ for the force

$\vec{F} = (x - z)\vec{i} + (1 - xy)\vec{j} + y\vec{k}$ as its point of application moves from the origin to the point A(1,1,1)
   (a) along the straight line x = y = z, and
   (b) along the curve
   $x = t^2, \quad y = t, \quad z = t^3, \quad 0 \le t \le 1.$

Solution.   (a) The integral to be evaluated is

$$\int_C \vec{F} \cdot d\vec{r} = \int_C (x - z)\,dx + (1 - xy)\,dy + y\,dz ,$$

which, for the straight line path x = y = z, becomes

$$\int_C \vec{F} \cdot d\vec{r} = \int_0^1 \frac{(1 - y^2 + y)}{}\,dy = \frac{7}{6} .$$

(b) Along the curve $x = t^2, \quad y = t, \quad z = t^3, \quad 0 \le t \le 1$ we get

$$\int_C \vec{F} \cdot d\vec{r} = \int_0^1 \frac{(t^2 - t^3) + (1 - t^3) + t}{}\,dt$$

$$= \int_0^1 \left(1 + 4t^3 - 2t^4\right) dt = \underline{\qquad} .$$

---

6. work, force   7. $\int_C$ M dx + N dy + P dz

8. (a) $\int_0^1 (1 + y - y^2)\,dy, \quad \frac{7}{6}$ (b) $\int_0^1 (t^2 - t^3)2t\,dt + (1 - t^3)\,dt + t \cdot 3t^2\,dt, \quad \frac{8}{5}$

Notice that the line integral depends on the path C joining the origin to A(1,1,1) since the values obtained in parts (a) and (b) are different.

9.   Suppose a particle moves upward along the helix whose vector equation is given by

$$\vec{r}(t) = \cos t\,\vec{i} + \sin t\,\vec{j} + t\vec{k}, \quad 0 \le t \le 2\pi .$$

Find the work done on the particle by the force

$$\vec{F} = (-zy)\vec{i} + (zx)\vec{j} + (xy)\vec{k}$$

as the point of application moves along C from the point A(1,0,0) to the point B(1,0,2π).

Solution. The work done is given by the integral

$$W = \int_C \vec{F} \cdot d\vec{r} = \int_a^b \vec{F}\big(x(t),y(t),z(t)\big) \cdot \frac{d\vec{r}}{dt}\, dt .$$

Now, in terms of the parameter t, the force $\vec{F}$ with point of application on the helix C is given by

$$\vec{F}\big(x(t),y(t),z(t)\big) = \underline{(-t\sin t)i + (t\cos t)j + (\sin t\cos t)k}$$

Also, $\dfrac{d\vec{r}}{dt} = (-\sin t)\vec{i} + \underline{\cos t}\,\vec{j} + \vec{k}$ so that

$$\vec{F} \cdot \frac{d\vec{r}}{dt} = t\sin^2 t + \underline{t\cos^2 t} = \underline{\sin t\cos t} .$$

As t varies from t = 0 to t = 2π, the point P(x,y,z) on C varies from $\underline{1,0,0}$ to $\underline{1,0,2\pi}$. Therefore, the work done by $\vec{F}$ from A to B is given by,

$$W = \int \underline{\qquad\qquad}\, dt = \underline{\qquad\qquad}\Big]\underline{\ } = 2\pi^2 .$$

10. Find the work done by the force

$$\vec{F} = y\vec{i} + z^2\vec{j} + x\vec{k} ,$$

as the point of application moves from A(0,0,0) along the y-axis to (0,-5,0), then in a straight line to the point B(0,1,1).

Solution. Let C_1 denote the line segment from A(0,0,0) to (0,-5,0), and let C_2 denote the straight line segment from (0,-5,0) to B(0,1,1). The path is sketched at the right. The total work done is the sum of two line integrals,

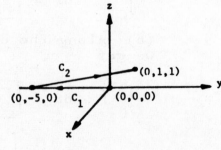

$$W = \int_{C_2} \vec{F} \cdot d\vec{r} + \int_{C_2} \vec{F} \cdot d\vec{r} .$$

---

9. $(-t\sin t)\vec{i} + (t\cos t)\vec{j} + (\sin t\cos t)\vec{k}$, $\cos t$, $t\cos^2 t + \sin t\cos t$, $t + \sin t\cos t$,   A(1,0,0), B(1,0,2π),

$$\int_0^{2\pi} (t + \sin t\cos t)\, dt, \quad \tfrac{1}{2}t^2 + \tfrac{1}{2}\sin^2 t\big]_0^{2\pi}$$

We want to find parametric representations for C_1 and C_2. The line segment C_1 is given by $x = 0$, $y = -t$, $z = 0$ for $0 \le t \le 5$. Notice that as t varies from $t = 0$ to $t = 5$, we move from the point $(0,0,0)$ to the point $(0,-5,0)$. A parametric representation for C_2 is given by $x = 0$, $y = 1 + 6t$, $z = \underline{1+t}$. As t varies from $t = \underline{-1}$ to $t = \underline{0}$, we move from the point $(0,-5,0)$ to the point $(0,1,1)$. Thus,

on C_1: $\vec{r} = \vec{r}_1(t) = -t\vec{j}$, $\dfrac{d\vec{r}_1}{dt} = \underline{-\vec{j}}$,

$\vec{F} = \vec{F}_1(t) = \underline{-t\vec{i}}$,   and $\vec{F}_1 \cdot \dfrac{d\vec{r}_1}{dt} = 0$;

on C_2: $\vec{r} = \vec{r}_2(t) = (1 + 6t)\vec{j} + \underline{(1+t)}\,\vec{k}$, $\dfrac{d\vec{r}_2}{dt} = \underline{6\vec{j} + \vec{k}}$,

$\vec{F} = \vec{F}_2(t) = \underline{(1+6t)\vec{i} + (1+t)^2\vec{j}}$,   and

$\vec{F}_2 \cdot \dfrac{d\vec{r}_2}{dt} = \underline{6(1+t)^2}$.

Therefore, the work done is

$$W = \int_0^5 \underline{\quad 0 \quad}\ dt + \int_{\underline{-1}}^{\underline{0}} 6(1 + t)^2\,dt = \underline{\quad 2 \quad} .$$

11.   Suppose $\vec{F}(x,y,z)$ is the velocity field of a fluid flowing through a region in space and that

$$C: \quad \vec{r}(t) = x(t)\vec{i} + y(t)\vec{j} + z(t)\vec{k}, \quad a \le t \le b,$$

is a curve lying in the region. Then the integral

$$\int_C \vec{F} \cdot d\vec{r}$$

is called the $\underline{\text{flow integral}}$ of $\vec{F}$ along C. If C is a closed curve, the value of the flow integral is called the $\underline{\text{circulation}}$ around the curve.

OBJECTIVE C: Given a continuous two-dimensional vector flow field $\vec{F}$, find the flux of $\vec{F}$ across a specified curve C in the xy-plane. Assume that C is piecewise smooth enough to have a tangent and that the direction across C is specified.

12.   If $\vec{n}$ is a unit normal vector indicating the direction of flow across C, then the flux of $\vec{F}$ across C is given as a line integral

$$\underline{\int_C \vec{F} \cdot \vec{n}\,ds} .$$

---

10. $(0,0,0)$ $(0,-5,0)$, $1 + t$, -1, 0, $-\vec{j}$, $-t\vec{i}$, $1 + t$, $6\vec{j} + \vec{k}$, $(1 + 6t)\vec{i} + (1 + t)^2\vec{j} + 0\vec{k}$,

$6(1 + t)^2$, 0, $\displaystyle\int_{-1}^{0}$, 2 11.   flow integral,   circulation   12. $\displaystyle\int_C \vec{F} \cdot \vec{n}\ ds$

13.   If C is a simple closed curve in the xy-plane, and if the
counterclockwise direction on C is taken as the positive
direction (as the direction of increasing arc length), then the
<u>flux</u> of the field $\vec{F}(x,y) = M(x,y)\vec{i} + N(x,y)\vec{j}$ outward across
C can be given by the line integral

$$\underline{\quad \oint_C M dy - N dx \quad}.$$

This integral has the advantage that it can be evaluated using
___<u>any</u>___ parameterization of C, provided we integrate in the
positive direction along C.

14.   Find the flux of the field $\vec{F} = xy^2\vec{i} + x^2y\vec{j}$ outward across the
circle $x^2 + y^2 = a^2$.

<u>Solution</u>.   A parameterization of the circle is $x = a \cos t$,
$y =$ _____, $0 \le t \le$ _____.   Using the line integral in
Problem 13 for the flux we have,

$$\oint_C (M\ dy - N\ dx) = \oint_C \underline{\qquad\qquad}$$

$$= \int_0^{2\pi} \underline{\qquad\qquad}\ dt$$

$$= 2a^4 \int_0^{2\pi} \tfrac{1}{4}(1 + \cos 2t) \underline{\qquad}\ dt$$

$$= 2a^4 \int_0^{2\pi} \tfrac{1}{4}(1 - \cos^2 2t)\ dt$$

$$= 2a^4 \int_0^{2\pi} \tfrac{1}{4}[1 - \tfrac{1}{2}(1 + \cos 4t)]\ dt$$

$$= 2a^4 \int_0^{2\pi} \underline{\qquad}\ dt$$

$$= 2a^4 [\tfrac{t}{8} - \tfrac{1}{32} \sin 4t]_0^{2\pi} = \underline{\qquad}.$$

The positive answer means that the net flux is _____.

15.3   GREEN'S THEOREM IN THE PLANE

15.   In mathematics, the <u>flux density</u> is called the <u>divergence</u> of
a vector flow field $\vec{F}$.   If $\vec{F}(x,y) = M(x,y)\vec{i} + N(x,y)\vec{j}$, then
div $\vec{F} =$ ___ $\frac{\partial M}{\partial x} + \frac{\partial N}{\partial y}$ ___ .

---

13. $\oint_C$ (M dy - N dx), any

14. a sin t, 2π, $xy^2\,dy - x^2y\,dx$, $2a^4 \cos^2 t \sin^2 t$, $1 - \cos 2t$, $\tfrac{1}{8} - \tfrac{1}{8} \cos 4t$, $\frac{\pi a^4}{2}$, outward

15. divergence, $\frac{\partial M}{\partial x} + \frac{\partial N}{\partial y}$

16. The divergence of a vector field $\vec{F}(x,y)$ is a _scalar_-valued function.

[OBJECTIVE A]: Find the divergence of a given vector field $\vec{F}(x,y)$.

17. For the vector field $\vec{F}(x,y) = (e^x + y^2)\vec{i} + (x^2y - y)\vec{j}$ the divergence is given by

$$\text{div } \vec{F} = \frac{\partial M}{\partial x} + \frac{\partial N}{\partial y}$$

$$= \underline{e^x} + x^2 - 1$$

$$= \underline{\hspace{3cm}} .$$

18. In mathematics, the circulation density is called the _curl_ of a vector flow field $\vec{F}$. If $\vec{F}(x,y) = M(x,y)\vec{i} + N(x,y)\vec{j}$, then

$$\text{curl } \vec{F} = \underline{\frac{dN}{dx} - \frac{dM}{dy}} .$$

19. The curl of a vector field $\vec{F}(x,y)$ is a _scalar_-valued function.

[OBJECTIVE B]: Find the curl of a given vector field $\vec{F}(x,y)$.

20. For the vector field in Problem 17, the curl is given by

$$\text{curl } \vec{F} = \frac{\partial N}{\partial x} - \frac{\partial M}{\partial y}$$

$$= 2xy - \underline{2y}$$

$$= \underline{2y(x-1)} .$$

21. In the flux-divergence form, Green's theorem is

$$\oint_C M \, dy - N \, dx = \iint_R \underline{\frac{dM}{dx} + \frac{dN}{dy}} \, dx \, dy.$$

That is, under suitable conditions the _flux_ of $\vec{F} = M\vec{i} + N\vec{j}$ outward across a simple closed curve C equals the double integral of the _divergence_ over the region R enclosed by C.

22. In the circulation curl form, Green's theorem is

$$\oint_C M \, dx + N \, dy = \iint_R \underline{\frac{dN}{dx} - \frac{dM}{dy}} \, dx \, dy.$$

---

16. scalar   17. e^x, $e^x + x^2 - 1$ 18. curl, $\frac{\partial N}{\partial x} - \frac{\partial M}{\partial y}$ 19. scalar

20. $2y$, $2y(x - 1)$ 21. $\frac{\partial M}{\partial x} + \frac{\partial N}{\partial y}$, flux, divergence of $\vec{F}$

That is, the counterclockwise _circulation_ of $\vec{F} = M\vec{i} + N\vec{j}$ around C equals the double integral of the _curl of F_ over the region R enclosed by C.

OBJECTIVE C: Use Green's theorem to evaluate a given line integral around a simple closed curve C. Assume all hypotheses of the theorem are satisfied, and that the curve C is oriented in the counterclockwise direction in the xy-plane.

23. If C is a suitable simple _closed_ curve in the xy-plane containing the region R as its interior, and if M, N, $\frac{\partial M}{\partial y}$, and $\frac{\partial N}{\partial x}$ are _continuous_ functions of (x,y) in R and on C, then

$$\oint_C (M\ dx + N\ dy) = \iint \frac{dN}{dx} - \frac{dM}{dy}\ dx dy$$

24. Give some examples of "suitable" regions R with a piecewise smooth boundary C in the xy-plane for which Green's theorem applies.

(a) _____

_____,

(b) _____,

(c) _____,

(d) _____.

25. Let C be the closed curve bounded by $x = 0$, $y = \sqrt{1 - x^2}$ and $y = 0$, oriented counter-clockwise. Evaluate the line integral

$$\oint_C 3y^4\ dx + (x^2 + 12y^3x)\ dy$$

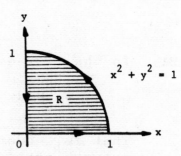

Solution. We will use Green's theorem. Thus, for $M = 3y^4$ and $N = x^2 + 12y^3x$,

$\frac{\partial N}{\partial x} - \frac{\partial M}{\partial y} = ($_____$) - 12y^3 =$ _____. Hence,

---

22. $\frac{\partial N}{\partial x} - \frac{\partial M}{\partial y}$, circulation, curl of $\vec{F}$ 23. closed, continuous, $\iint_R [\frac{\partial N}{\partial x} - \frac{\partial M}{\partial y}]\ dx\ dy$

24. (a) R is bounded by a simple closed curve C such that a line parallel to either axis cuts C in at most two points
(b) R is a rectangle   (c) R is an annulus
(d) R combines regions like those in (a), (b), or (c)

$$\oint_C 3y^4 \, dx + (x^2 + 12y^3x) \, dy = \int \underline{\quad} \int \underline{\quad} \underline{\hspace{3cm}} \, dy \, dx$$

$$= \int_0^1 \underline{\hspace{3cm}} \, dx = \underline{\hspace{2cm}} \, .$$

26. Let us evaluate the line integral

$$\oint_C xy \, dx + \left(x^{3/2} + y^{3/2}\right) dy \, ,$$

where C is the triangle with vertices (0,0), (1,0), (1,1) oriented counterclockwise. Using Green's theorem,

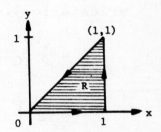

$$\oint_C xy \, dx + \left(x^{3/2} + y^{3/2}\right) dy$$

$$= \iint_R \underline{\hspace{2cm}} \, dA = \int \underline{\quad} \int \underline{\quad} \left(\tfrac{3}{2}x^{1/2} - x\right) dy \, dx$$

$$= \int_0^1 \underline{\hspace{3cm}} \, dx = \tfrac{3}{5}x^{5/2} - \underline{\hspace{2cm}}]_0^1 = \underline{\hspace{2cm}} \, .$$

[OBJECTIVE D]: Use Green's theorem to find the area of a given suitable region R in the xy-plane with a piecewise smooth boundary C.

27. For regions R, like those specified in Problem 24 above, the area of R can be calculated via Green's theorem as the line integral <u>   </u>. (See Problems 19-22 on page 966 of Finney/Thomas.)

28. Let us find the area of the region bounded below by the x-axis and above by one arch of the cycloid parameterized by

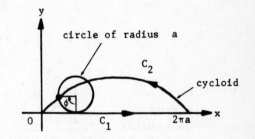

$$x = a(\phi - \sin \phi),$$

$$y = a(1 - \cos \phi),$$

$0 \leq \phi \leq 2\pi$. (See the figure at the right.) We take the counterclockwise orientation on the curve C bounding the region. The curve C is made up of two smooth curves: the x-axis C_1, and the cycloid arch C_2. Now,

---

25. $2x + 12y^3$, $2x$, $\displaystyle\int_0^1 \int_0^{\sqrt{1-x^2}} 2x \, dy \, dx$, $2x\sqrt{1-x^2}$, $\tfrac{2}{3}$

26. $\tfrac{3}{2}x^{1/2} - x$, $\displaystyle\int_0^1 \int_0^x$, $x\left(\tfrac{3}{2}x^{1/2} - x\right)$, $\tfrac{1}{3}x^3$, $\tfrac{4}{15}$ 27. $\tfrac{1}{2}\oint_C x \, dy - y \, dx$

on C_1: $x \, dy - y \, dx = \underline{\hspace{3cm}} = 0,$ and

on C_2: $x \, dy - y \, dx = a(\phi - \sin \phi)(a \sin \phi \, d\phi)$

$$- \underline{\hspace{5cm}}$$

$$= \underline{\hspace{3cm}} .$$

Therefore, the area of the region is given by (use the correct orientation),

$$A = \frac{1}{2} \oint_C x \, dy - y \, dx$$

$$= \frac{a^2}{2} \int \underline{\hspace{0.8cm}} \underline{\hspace{4cm}} \, d\phi$$

$$= \frac{a^2}{2} \left[\sin \phi - \cos \phi + \underline{\hspace{3cm}} \right]_{2\pi}^{0} = \underline{\hspace{3cm}} .$$

15.4   SURFACE AREA AND SURFACE INTEGRALS

OBJECTIVE A :   Using integration, find the area of a specified surface $f(x,y,z) = c$ in space.

29.   Suppose $f(x,y,z) = c$ is a surface in space. Let   R   denote the region representing the "shadow" of the surface over some plane whose normal is the vector $\vec{p}$.   Let γ be the angle between $\vec{\nabla}f$ and $\vec{p}$.   Then, assuming $\vec{\nabla}f$ is not parallel to the plane, so that $\vec{\nabla}f \cdot \vec{p} \neq 0$,   the area of the surface over R   is given by

$$\text{Surface area} = \iint_R = \frac{1}{|\cos \gamma|} \, dA$$

$$= \iint_R \underline{\hspace{4cm}} \, dA .$$

30.   Let us find the area of the surface of that portion of the sphere $x^2 + y^2 + z^2 = a^2$ that lies inside the elliptic cylinder $4y^2 + x^2 = a^2$.   The figure at the top of the next page shows that portion which lies above the first quadrant of the xy-plane. Because of the symmetries of the ellipse and the sphere, the total surface area is equal to   8   times that shown in the figure. The gradient vector $\vec{\nabla}f = \underline{\hspace{3cm}}$ is normal to the surface of the sphere, and the unit vector $\vec{k}$ is

---

28. $x \cdot 0 - 0 \cdot dx,$ $a(1 - \cos \phi)(a - a \cos \phi) \, d\phi,$ $a^2(\phi \sin \phi + 2 \cos \phi - 2),$

$\displaystyle\int_{2\pi}^{0} (\phi \sin \phi + 2 \cos \phi - 2) \, d\phi,$ $2 \sin \phi - 2\phi,$ $3\pi a^2$

29. $\dfrac{|\vec{\nabla}f|}{|\vec{\nabla}f \cdot \vec{p}|}$

perpendicular to the ground plane.
The region of integration R is
described by the inequalities
$0 \leq x \leq a$ and
_____ $< y \leq$ _____.
Therefore, the total surface area
is given by

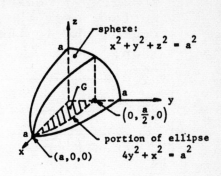

sphere:
$x^2 + y^2 + z^2 = a^2$

$\left(0, \frac{a}{2}, 0\right)$

portion of ellipse
$4y^2 + x^2 = a^2$

$(a,0,0)$

$$S = 8 \iint_R \frac{|\vec{\nabla} f|}{|\vec{\nabla} f \cdot \vec{k}|} \, dA$$

$$= 8 \iint_R \underline{\hspace{3cm}} \, dA$$

$$= 8 \int_{\underline{\hspace{0.5cm}}}^{\underline{\hspace{0.5cm}}} \int_{\underline{\hspace{0.5cm}}}^{\underline{\hspace{0.5cm}}} \frac{a}{\sqrt{a^2 - x^2 - y^2}} \, dy \, dx$$

$$(\text{since} \quad x^2 + y^2 + z^2 = a^2, \quad z \geq 0)$$

$$= 8a \int_0^a \left. \sin^{-1} \frac{y}{\sqrt{a^2 - x^2}} \right]_{y=0}^{(a^2-x^2)^{1/2}/2} \, dx$$

$$= 8a \int_0^a \underline{\hspace{2cm}} \, dx = \underline{\hspace{2cm}} \, .$$

31. If the surface is given in the form $z = g(x,y)$, we define the
function $f(x,y) = g(x,y) - z = 0$. In this situation
$$\vec{\nabla} f = \underline{\hspace{4cm}} \, .$$
Let R denote the region in the xy-plane representing the
"shadow" of the surface $z = g(x,y)$. From Problem 29, the
surface area is given by

$$\text{Surface area} = \iint_R \frac{|\vec{\nabla} f|}{|\vec{\nabla} f \cdot \vec{k}|} \, dA$$

$$= \iint_R \underline{\hspace{4cm}} \, dx \, dy \, .$$

(See Problems 31-36 on page 977 of Finney/Thomas.)

32. Suppose we wish to find the surface area of that part of the
cylinder $z^2 + x^2 = a^2$, $a > 0$, which lies above the triangle
in the xy-plane bounded by $x = 0$, $y = 0$, and $x + y = a$. A
sketch of the surface is shown at the top of the next page.
Since $z \geq 0$, we can take $z = \sqrt{a^2 - x^2}$ as the function

---

30. $2x\vec{i} + 2y\vec{j} + 2z\vec{k}$, $0 \leq y \leq \frac{1}{2}\sqrt{a^2 - x^2}$, $\dfrac{\sqrt{4x^2 + 4y^2 + 4z^2}}{2z}$, $\displaystyle\int_0^a \int_0^{\sqrt{a^2-x^2}/2}$, $\dfrac{\pi}{6}$, $\dfrac{4\pi a^2}{3}$

31. $\vec{\nabla} f = \dfrac{\partial g}{\partial x}\vec{i} + \dfrac{\partial g}{\partial y}\vec{j} - \vec{k}$, $\sqrt{g_x^2 + g_y^2 + 1}$

z = g(x,y)   describing the
surface over the triangular
region R in the xy-plane.
Then,

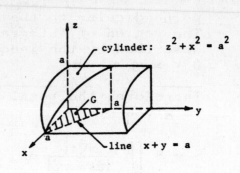

cylinder: $z^2 + x^2 = a^2$

G

line x + y = a

$\dfrac{\partial z}{\partial x}$ = _____ and

$\dfrac{\partial z}{\partial y}$ = _____ . Thus,

$$\sqrt{\left(\dfrac{\partial z}{\partial x}\right)^2 + \left(\dfrac{\partial z}{\partial y}\right)^2 + 1}$$

= _____ .

The region of integration
R is described by the
inequalities _____ ≤ x ≤ _____ and _____ ≤ y ≤ _____.
Therefore, from Problem 31, the surface area is given by the
integral

$$\text{Surface area} = \int_{\overline{}}^{\overline{}} \int_{\overline{}}^{\overline{}} \underline{\hspace{4cm}}\ dy\ dx$$

$$= a \int_0^a \dfrac{a - x}{\sqrt{a^2 - x^2}}\ dx$$

$$= a^2 \int_0^a \dfrac{dx}{\sqrt{a^2 - x^2}} - a \int_0^a \dfrac{x\ dx}{\sqrt{a^2 - x^2}}$$

$$= a^2 \left(\underline{\hspace{2cm}}\right) + a\sqrt{a^2 - x^2}\,\Big]_{x=0}^{x=a} = \underline{\hspace{3cm}} .$$

OBJECTIVE B :   Let S denote a specified surface in space. Assuming
that g(x,y,z) is a given continuous function on S,
find the surface integral

$$\iint\limits_S g(x,y,z)\ d\sigma.$$

33. Consider the surface integral
$\iint\limits_S z^2\ d\sigma$,   where S is the

portion of the cone $z^2 = x^2 + y^2$
between the planes z = 2 and
z = 5. The surface S and its
projection R onto the xy-plane
are sketched at the right.
Observe that R is the ring
_____ ≤ $x^2 + y^2$ ≤ _____. The
function z = _____ defines
the surface S over R.

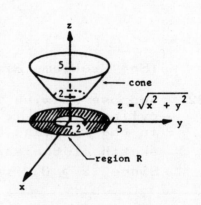

cone

$z = \sqrt{x^2 + y^2}$

region R

---

32. $\dfrac{-x}{\sqrt{a^2 - x^2}}$, 0, $\dfrac{a}{\sqrt{a^2 - x^2}}$, $0 \le x \le a,\ 0 \le y \le a - x$, $\displaystyle\int_0^a \int_0^{a-x} \dfrac{a}{\sqrt{a^2 - x^2}}\ dy\ dx$, $\sin^{-1}\dfrac{x}{a}$, $\dfrac{a^2}{2}(\pi - 2)$

Thus, from Problem 31, we use the gradient

$$\vec{\nabla}z = \frac{\partial z}{\partial x}\vec{i} + \frac{\partial z}{\partial y}\vec{j} - \vec{k} \ .$$

Now,

$$\frac{\partial z}{\partial x} = \underline{\hspace{2cm}} = \frac{x}{z} \quad \text{and} \quad \frac{\partial z}{\partial y} = \underline{\hspace{1.5cm}} = \underline{\hspace{1.5cm}} \ .$$

Hence,

$$d\sigma = \sqrt{1 + \left(\frac{x^2}{z^2}\right) + (\underline{\hspace{1.5cm}})} \ dA = \underline{\hspace{1.5cm}} \ dA$$

because $x^2 + y^2 + z^2 = 2z^2$ on S. Therefore

$$\iint\limits_{S} z^2 \ d\sigma = \iint\limits_{R} (x^2 + y^2)\sqrt{2} \ dA = \int_{\underline{\ }}^{\overline{\ }} \int_{\underline{\ }}^{\overline{\ }} \underline{\hspace{3cm}} \ dr \ d\theta$$

$$= \underline{\hspace{2cm}} \ (5^4 - 2^4) = \frac{609\pi\sqrt{2}}{2} \approx 1352.86 \ .$$

OBJECTIVE C : Evaluate the flux integral

$$\iint\limits_{S} \vec{F} \cdot \vec{n} \ d\sigma \ ,$$

for a specified vector field $\vec{F}$ and surface S, where $\vec{n}$ is a unit vector normal to S and pointing in a consistent way (e.g., always outward or always inward). Assume that the components of $\vec{F}$ are continuous on S.

34. Let S be the paraboloid $z = 1 - x^2 - y^2$ above the xy-plane, and let $\vec{n}$ be a unit normal to S at each point, which is directed upward. Let $\vec{F}$ be the vector field on S given by $\vec{F} = x\vec{i} + y\vec{j} + z\vec{k}$. We want to find the surface integral

$$\iint\limits_{S} \vec{F} \cdot \vec{n} \ d\sigma \ .$$

Let us first calculate the vector $\vec{n}$. If we define $g(x,y,z) = z - (1 - x^2 - y^2)$, then the gradient vector $\vec{\nabla}g = 2x\vec{i} + \underline{\hspace{1.5cm}}\vec{j} + \vec{k}$ is perpendicular to the surface S at each point. Moreover, it is directed upward and outward so it is pointing in the direction specified for $\vec{n}$. The unit normal $\vec{n}$ can be written as

$$\vec{n} = \frac{\vec{\nabla}g}{|\vec{\nabla}g|} = \underline{\hspace{4cm}} \ .$$

Therefore,

$$\vec{F} \cdot \vec{n} = \underline{\hspace{3cm}} = \frac{1 + x^2 + y^2}{\sqrt{4x^2 + 4y^2 + 1}}$$

---

33. $4 \leq x^2 + y^2 \leq 25, \ \sqrt{x^2 + y^2}, \ \dfrac{x}{\sqrt{x^2 + y^2}}, \ \dfrac{y}{\sqrt{x^2 + y^2}}, \ \dfrac{y}{z}, \ \dfrac{y^2}{z^2}, \ \sqrt{2}, \ \displaystyle\int_0^{2\pi}\int_2^5 \sqrt{2}\,r^3 \ dr \ d\theta, \ \dfrac{\sqrt{2}\pi}{2}$

because $z = 1 - x^2 - y^2$ on the surface   S.   Next,

$$d\sigma = \sqrt{1 + z_x^2 + z_y^2} \; dA = \underline{\hspace{2cm}} \; dA.$$

The surface   S   projects onto the region   R   in the xy-plane described by $\underline{\hspace{2cm}}$.   Therefore

$$\iint\limits_{S} \vec{F} \cdot \vec{n} \; d\sigma = \iint\limits_{R} (\underline{\hspace{3cm}}) \; dA$$

$$= \int \underline{\hspace{0.5cm}} \int \underline{\hspace{0.5cm}} \underline{\hspace{3cm}} \; dr \; d\theta$$

$$\underline{\uparrow\hspace{0.8cm}} \quad \text{in polar coordinates}$$

$$= \underline{\hspace{1.5cm}} \approx 4.71.$$

15.5   THE DIVERGENCE THEOREM

OBJECTIVE A :   Given a vector field $\vec{F} = M\vec{i} + N\vec{j} + P\vec{k}$ with suitable continuity conditions, verify the divergence theorem

$$\iint\limits_{S} \vec{F} \cdot \vec{n} \; d\sigma = \iiint\limits_{D} \text{div } \vec{F} \; dV \; .$$

Assume that   D   is a convex region with no holes or bubbles enclosed by a piecewise smooth surface   S   with outward normal $\vec{n}$.

35.   Let us verify the divergence theorem for the field $\vec{F} = (2xyz)\vec{i} + (x^2 z)\vec{j} + (x^2 y + 1)\vec{k}$ and the cube with center at the origin and faces in the planes $x = \pm 1$, $y = \pm 1$, $z = \pm 1$. First,

$$\text{div } \vec{F} = \frac{\partial M}{\partial x} + \frac{\partial N}{\partial y} + \frac{\partial P}{\partial z} = 2yz + \underline{\hspace{1.5cm}} + 0 = \underline{\hspace{1.5cm}} \; .$$

Thus, if   D   is the region enclosed by the cube,

$$\iiint\limits_{D} \text{div } \vec{F} \; dV = \int_{-1}^{1} \int_{-1}^{1} \int_{-1}^{1} 2yz \; dz \; dy \; dx$$

$$= \int_{-1}^{1} \int_{-1}^{1} \underline{\hspace{2cm}} \Big]_{z=-1}^{z=1} dy \; dx = \underline{\hspace{1.5cm}} \; .$$

Next,   we compute $\iint \vec{F} \cdot \vec{n} \; d\sigma$ as the sum of the integrals over the six faces separately.   We begin with the two faces perpendicular to the x-axis.   For the face $x = -1$ and the face $x = +1$,   respectively, we have the first and second lines of the following table (complete the table):

---

34.   2y, $\dfrac{2x\vec{i} + 2y\vec{j} + \vec{k}}{\sqrt{4x^2 + 4y^2 + 1}}$, $\dfrac{2x^2 + 2y^2 + z}{\sqrt{4x^2 + 4y^2 + 1}}$, $\sqrt{4x^2 + 4y^2 + 1}$, $0 \leq x^2 + y^2 \leq 1$, $1 + x^2 + y^2$,

$\int_{0}^{2\pi} \int_{0}^{1} (1 + r^2) \, r \; dr \; d\theta$, $\dfrac{3\pi}{2}$

range of integration	outward unit normal	field $\vec{F}$
$-1 \leq y \leq 1, \; -1 \leq z \leq 1$	$-\vec{i}$	$(-2yz)\vec{i} + z\vec{j} + (y+1)\vec{k}$

For each of these planes, $x = x(y,z)$ is a constant function of (y,z) so that

$$d\sigma = \sqrt{1 + \left(\frac{\partial x}{\partial y}\right)^2 + \left(\frac{\partial x}{\partial z}\right)^2} \; dy \; dz = \underline{\hspace{2cm}} .$$

Therefore, the sum of the integrals over these two faces is

$$\iint \vec{F} \cdot \vec{n} \; d\sigma = \int_{-1}^{1} \int_{-1}^{1} 2yz \; dz \; dy + \underline{\hspace{4cm}}$$

$$= 2 \int_{-1}^{1} yz^2 \big]_{z=-1}^{z=1} \; dy = \underline{\hspace{2cm}} .$$

Now consider the two faces perpendicular to the y-axis. For the face $y = -1$ and the face $y = +1$, respectively, complete the first and second lines of the following table:

range of integration	outward unit normal	field $\vec{F}$
$-1 \leq x \leq 1, \; -1 \leq z \leq 1$	$-\vec{j}$	$(-2xz)\vec{i} + (x^2z)\vec{j} + (-x^2+1)\vec{k}$

For each of these planes $d\sigma = \underline{\hspace{2cm}}$. Therefore, the sum of the surface integrals over these two faces is

$$\iint \vec{F} \cdot \vec{n} \; d\sigma = \int_{-1}^{1} \int_{-1}^{1} (-x^2z) \; dz \; dx + \underline{\hspace{3cm}}$$

$$= \underline{\hspace{2cm}} .$$

Finally, we take the two faces perpendicular to the z-axis. For the face $z = -1$ and the face $z = +1$, respectively, we obtain the first and second lines of the following table:

range of integration	outward unit normal	field $\vec{F}$
$-1 \leq x \leq 1, \; -1 \leq y \leq 1$	$-\vec{k}$	$(-2xy)\vec{i} + (-x^2)\vec{j} + (x^2y+1)\vec{k}$
$-1 \leq x \leq 1, \; -1 \leq y \leq 1$	$\vec{k}$	$(2xy)\vec{i} + x^2\vec{j} + (x^2y + 1)\vec{k}$

---

35. $0, \; 2yz, \; yz^2, \; 0, \; -1 \leq y \leq 1, \; -1 \leq z \leq 1, \; \vec{i}, \; (2yz)\vec{i} + z\vec{j} + (y+1)\vec{k}, \;$ dy dz,

$\int_{-1}^{1} \int_{-1}^{1} 2yz \; dz \; dy, \; 0, \; -1 \leq x \leq 1, \; -1 \leq z \leq 1, \; \vec{j}, \; (2xz)\vec{i} + (x^2z)\vec{j} + (x^2 + 1)\vec{k},$

dz dx, $\int_{-1}^{1} \int_{-1}^{1} x^2z \; dz \; dx, \; 0, \; 0$

For each of these planes $d\sigma = dx\ dy$,   and the sum of the surface integrals over these two faces is

$$\iint \vec{F} \cdot \vec{n}\ d\sigma = \int_{-1}^{1} \int_{-1}^{1} -(x^2 y + 1)\ dx\ dy + \int_{-1}^{1} \int_{-1}^{1} (x^2 y + 1)\ dx\ dy$$

$$= 0 \ .$$

Therefore, the surface integral over the six faces of the cube is

$$\iiint_{D} \text{div } \vec{F}\ dV = \underline{\hspace{2cm}} = \iint_{S} \vec{F} \cdot \vec{n}\ d\sigma \ .$$

36.   Give some examples of "suitable" regions   D   enclosed by a piecewise smooth bounding surface   S   for which the divergence theorem applies.   One example is when   D   is a convex region with no holes such that when   D   is projected into a coordinate plane it produces a simply connected region   R,   with the property that any line perpendicular to that coordinate plane at an interior point of   R   intersects the bounding surface   S   in at most _____ points.   Other examples are:

   (a)   _____ ,

   (b)   _____ ,

   (c)   _____
         _____ .

37.   Let us verify the divergence for the vector field
   $\vec{F} = x\vec{i} + y\vec{j} + z\vec{k}$,   where   D   is the region inside the cylinder
   $x^2 + y^2 = 1$ and between the planes $z = 0$ and $z = 2$.   Now,
   $$\text{div } \vec{F} = \frac{\partial M}{\partial x} + \frac{\partial N}{\partial y} + \frac{\partial P}{\partial z} = \underline{\hspace{2cm}} \ ,$$
   so that

$$\iiint_{D} \text{div } \vec{F}\ dV = 3 \iiint_{D} dV = \underline{\hspace{2.5cm}} = 6\pi \ .$$

The surface integral $\iint \vec{F} \cdot \vec{n}\ d\sigma$ is the sum of three integrals:

one over the top, another one over the bottom, and a third one over the side of the cylinder.   We calculate each of these separately.

The top: $z = 2$, $0 \le x^2 + y^2 \le 1$, $\vec{n} = \underline{\hspace{2cm}}$ is the outer normal, and

$$\vec{F} = \underline{\hspace{2.5cm}} \ .$$

---

36.   two,   (a) cubes
      (b) two concentric spheres
      (c) regions that can be split up into a finite number of simple regions of the type described above

Then,

$$\iint \vec{F} \cdot \vec{n} \; d\sigma = \iint \underline{\hspace{2cm}} \; d\sigma = 2 \times \text{area of top} = \underline{\hspace{2cm}} \; .$$

The bottom: $z = 0$, $0 \le x^2 + y^2 \le 1$, $\vec{n} = \underline{\hspace{2cm}}$ is the outer normal, and

$$\vec{F} = \underline{\hspace{3cm}} \; .$$

Then,

$$\iint \vec{F} \cdot \vec{n} \; d\sigma = \iint \underline{\hspace{2cm}} \; d\sigma = 0 \; .$$

The side: $x^2 + y^2 = 1$, $0 \le z \le 2$, $\vec{F} = \underline{\hspace{2cm}}$ and

$$\vec{n} = \frac{x\vec{i} + y\vec{j}}{\sqrt{x^2 + y^2}} = x\vec{i} + y\vec{j} \quad \text{is the outer normal. Hence,}$$

$$\iint \vec{F} \cdot n \; d\sigma = \iint \underline{\hspace{3cm}} \; d\sigma = \iint d\sigma = \underline{\hspace{3cm}} = 4\pi \; .$$

Therefore, the surface integral over the entire cylindrical can is

$$\iiint_D \text{div} \; \vec{F} \; dV = \underline{\hspace{2cm}} + 0 + 4\pi = \iint_S \vec{F} \cdot \vec{n} \; d\sigma \; .$$

OBJECTIVE B :   Given a continuous three-dimensional vector flow field $\vec{F}$, find the flux of $\vec{F}$ outward through a specified smooth closed surface S enclosing a region D.

38.   Consider the flow field $\vec{F} = y\vec{i} - x\vec{j} + (x^2 + y^2)\vec{k}$ through the portion of the paraboloid $z = 9 - x^2 - y^2$ above the xy-plane, and directed outward through the surface. We want to find the flux outward. Therefore, we seek to evaluate the surface integral

$$\underline{\hspace{5cm}} \; ,$$

where $\vec{n}$ is the outward-pointing unit normal to the paraboloid. Let us first calculate this vector $\vec{n}$. If we define $g(x,y,z) = z - (9 - x^2 - y^2)$, then the vector $\underline{\hspace{2cm}}$ is a unit vector normal to the surface at each point.

We find this vector to be $\dfrac{\vec{\nabla} g}{|\vec{\nabla} g|} = \dfrac{2x\vec{i} + 2y\vec{j} + \vec{k}}{\underline{\hspace{2cm}}}$, and it is directed upward from the surface of the paraboloid. Thus we can take $\vec{n} = \dfrac{\vec{\nabla} g}{|\vec{\nabla} g|}$. Then,

$$\vec{F} \cdot \vec{n} = \frac{\underline{\hspace{2cm}}}{\sqrt{4x^2 + 4y^2 + 1}} \; .$$

---

37. 3, $3\pi(1)^2(2)$, $\vec{k}$, $x\vec{i} + y\vec{j} + 2\vec{k}$, 2, 2π, $-\vec{k}$, $x\vec{i} + y\vec{j}$, 0, $x\vec{i} + y\vec{j} + z\vec{k}$, $x^2 + y^2$,

$2\pi(1)(2)$, 2π

Next,

$$d\sigma = \sqrt{1 + \left(\frac{\partial z}{\partial x}\right)^2 + \left(\frac{\partial z}{\partial y}\right)^2}\ dA = \underline{\hspace{3cm}}\ dx\ dy\ .$$

The paraboloid S projects onto the region R in the xy-plane bounded by the circle $x^2 + y^2 = 9$. Therefore, the flux of $\vec{F}$ is,

$$\iint_S \vec{F} \cdot \vec{n}\ d\sigma = \iint_R \left(\frac{x^2 + y^2}{\sqrt{4x^2 + 4y^2 + 1}}\right) \sqrt{4x^2 + 4y^2 + 1}\ dx\ dy$$

$$= \int \underline{\underline{\hspace{1cm}}} \int \underline{\underline{\hspace{1cm}}} \underline{\hspace{1cm}}\ dr\ d\theta = \frac{\pi 3^4}{2} \approx 127.23\ .$$

$\uparrow\_$ in polar coordinates

15.6 STOKES'S THEOREM

OBJECTIVE A : Given a vector field $\vec{F} = M\vec{i} + N\vec{j} + P\vec{k}$, where M, N, and P are continuous functions of (x,y,z), together with their first-order partial derivatives, throughout a region D containing a specified smooth, simply connected, orientable surface S bounded by a simple closed curve C, verify Stokes's theorem. Assume that the positive direction around C is the one induced by the positive orientation of S.

39.   Under the hypotheses stated in the above Objective, if $\vec{n}$ is a positive unit vector normal to S, then

$$\oint_C \vec{F} \cdot d\vec{r} = \underline{\hspace{4cm}}\ .$$

40.   Consider the oriented triangle shown in the figure at the right. Let
$$\vec{F} = y^2\vec{i} + xy\vec{j} - 2xz\vec{k}\ .$$
We will verify Stokes's theorem for the triangular plane bounded by the sides C_1, C_2, C_3 directed as shown. First, let us find the positive unit vector normal to the plane that agrees with the orientation around the triangle. A vector normal to the plane is
$$2\vec{i} \times (2\vec{j} + \vec{k}) = \underline{\hspace{3cm}}\ .$$
Notice that this vector points

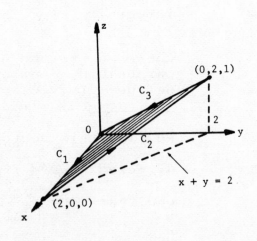

---

38. $\displaystyle\iint_S \vec{F} \cdot \vec{n}\ d\sigma,\ \ \frac{\vec{\nabla}g}{|\vec{\nabla}g|},\ \ \sqrt{4x^2 + 4y^2 + 1},\ \ 2xy - 2xy + x^2 + y^2,\ \ \sqrt{4x^2 + 4y^2 + 1},\ \ \int_0^{2\pi} \int_0^3 r^3\ dr\ d\theta$

39. $\displaystyle\iint_S (\text{curl }\vec{F}) \cdot \vec{n}\ d\sigma$

upward (positive $\vec{k}$ component) and toward the left in the figure (negative $\vec{j}$ component), so it points in the positive direction relative to the counterclockwise direction around the boundary. Therefore, the unit vector $\vec{n}$ is given by $\vec{n} = $ _____. An equation of the plane is $-2(y - 0) + 4(z - 0) = 0$, or $z = $ _____. Also,

$$\text{curl } \vec{F} = \begin{vmatrix} \vec{i} & \vec{j} & \vec{k} \\ \frac{\partial}{\partial x} & \frac{\partial}{\partial y} & \frac{\partial}{\partial z} \\ y^2 & xy & -2xz \end{vmatrix} = \underline{\hspace{4cm}} \; .$$

For the element of surface area we use,

$$d\sigma = \sqrt{1 + \left(\frac{\partial z}{\partial x}\right)^2 + \left(\frac{\partial z}{\partial y}\right)^2} \; dx \; dy = \underline{\hspace{3cm}} \; dx \; dy$$

or $d\sigma = \frac{1}{2}\sqrt{5} \; dx \; dy$. Therefore,

$$\text{curl } \vec{F} \cdot \vec{n} \; d\sigma = (2z\vec{j} - y\vec{k}) \cdot (-\tfrac{1}{2}\vec{j} + \vec{k})dx \; dy = (-z - y)dx \; dy$$

$$= \underline{\hspace{2.5cm}} \; dx \; dy \quad \text{on the plane.}$$

Hence,

$$\iint\limits_{S} \text{curl } \vec{F} \cdot \vec{n} \; d\sigma = \int \underline{\hspace{0.5cm}} \int \underline{\hspace{0.5cm}} -\tfrac{3}{2}y \; dy \; dx = \int_0^2 -\tfrac{3}{4}(2 - x)^2 \; dx$$

$$= \underline{\hspace{1.5cm}} \; .$$

Now, to find the line integral $\oint_C \vec{F} \cdot d\vec{r}$, we will

calculate three integrals, one for C_1, C_2, and C_3 separately, and sum them. The integrand in the line integral is

$$\vec{F} \cdot (dx\,\vec{i} + dy\,\vec{j} + dz\,\vec{k}) = \underline{\hspace{4cm}} \; .$$

On C_1: $z = 0$, $y = 0$, $0 \leq x \leq 2$, and thus

$$\vec{F} \cdot d\vec{r} = \underline{\hspace{1.5cm}} \; . \quad \text{Hence} \quad \int_{C_1} \vec{F} \cdot d\vec{r} = 0 \; .$$

On C_2: The line segment from $(2,0,0)$ to $(0,2,1)$ can be represented parametrically by $x = 2 - 2t$, $y = 2t$, $z = t$ for $0 \leq t \leq 1$. Then, $\vec{F} \cdot d\vec{r}$ on the segment can be expressed in terms of t by,

$$\vec{F} \cdot d\vec{r} = (2t)^2(-2 \; dt) + \underline{\hspace{2.5cm}} - 2(2 - 2t)t \; dt$$

$$= \underline{\hspace{2.5cm}} dt, \quad \text{and therefore}$$

$$\int_{C_2} \vec{F} \cdot d\vec{r} = \int \underline{\hspace{0.5cm}} 4(-3t^2 + t)dt = \underline{\hspace{1.5cm}} \; .$$

$\uparrow$ (check the orientation for C_2 here)

---

40. $-2\vec{j} + 4\vec{k}$, $\dfrac{-\vec{j} + 2\vec{k}}{\sqrt{5}}$, $\dfrac{y}{2}$, $2z\vec{j} - y\vec{k}$, $\sqrt{1 + 0 + \frac{1}{4}}$, $-\frac{3}{2}y$, $\displaystyle\int_0^2 \int_0^{2-x}$, -2, $y^2 \; dx + xy \; dy - 2xz \; dz$,

0, $(2 - 2t)(2t)(2 \; dt)$, $4(-3t^2 + t)$, $\displaystyle\int_0^1$, -2, 0, -2

On C_3: $x = 0$, $z = \frac{y}{2}$, $2 \geq y \geq 0$,   and thus

$$\vec{F} \cdot d\vec{r} = y^2 \cdot 0 + 0dy - 0dz = \underline{\hspace{2cm}}\text{ and}$$

$$\int_{C_3} \vec{F} \cdot d\vec{r} = 0 \ .$$

Therefore, summing the three line integrals,

$$\oint_C \vec{F} \cdot d\vec{r} = \underline{\hspace{2cm}} = \iint_S \text{curl } \vec{F} \cdot \vec{n} \ d\sigma \ .$$

41. Stokes's theorem says that, under conditions normally met in practice, the \underline{\hspace{3cm}} of a vector field $\vec{F}$ around the boundary of a surface in space, in a direction counterclockwise with respect to the \underline{\hspace{2cm}} vector $\vec{n}$ of the surface, is equal to the double integral of \underline{\hspace{2.5cm}} over the surface.

42. An important vector identity states that
$$\text{curl grad } f \quad \text{or} \quad \vec{\nabla} \times \vec{\nabla} f$$
equals \underline{\hspace{3cm}}. The identity holds for any function $f(x,y,z)$ having \underline{\hspace{6cm}}.

OBJECTIVE B: Use the surface integral in Stokes's theorem to calculate the circulation of a specified vector field $\vec{F}$ around a space curve C in an indicated direction.

43. Find the circulation of $\vec{F} = z\vec{i} - x\vec{j} + y\vec{k}$ around C: the circle $x^2 + y^2 = 4$ in the xy-plane, counterclockwise when viewed from above.

Solution. According to Stokes's theorem, the circulation is given by the integral

$$\iint_S \text{curl } \vec{F} \cdot \vec{n} \ d\sigma$$

where S is the hemisphere \underline{\hspace{3cm}}. For the curl of $\vec{F} = z\vec{i} - x\vec{j} + y\vec{k}$, we have

$$\vec{\nabla} \times \vec{F} = \begin{vmatrix} \vec{i} & \vec{j} & \vec{k} \\ \frac{\partial}{\partial x} & \frac{\partial}{\partial y} & \frac{\partial}{\partial z} \\ z & -x & y \end{vmatrix} = \underline{\hspace{4cm}} \ .$$

The outer unit normal of S is the vector

$$\vec{n} = \frac{x\vec{i} + y\vec{j} + z\vec{k}}{\sqrt{x^2 + y^2 + z^2}} = \underline{\hspace{4cm}} \ .$$

---

41. circulation, normal, curl $\vec{F} \cdot \vec{n}$

42. the zero vector,   continuous first and second partial derivatives

Next, we calculate the element of surface area. The gradient to the surface S satisfies

$$|\vec{\nabla}F| = |2x\vec{i} + 2y\vec{j} + 2z\vec{k}| = 2\sqrt{x^2 + y^2 + z^2} = 4 \ ,$$
$$|\vec{\nabla}F \cdot \vec{p}| = |\vec{\nabla}F \cdot \vec{k}| = |2z| = 2z \ ,$$

$$d\sigma = \frac{|\vec{\nabla}F|}{|\vec{\nabla}F \cdot \vec{p}|} \ dA = \frac{2}{z} \ dA \ .$$

(Note $F(x,y,z) = x^2 + y^2 + z^2 = 4$ is the <u>surface</u> S and <u>not</u> the vector field $\vec{F} = z\vec{i} - x\vec{i} + y\vec{k}$.) Thus,

$$\iint\limits_{S} \text{curl } \vec{F} \cdot \vec{n} \ d\sigma = \iint\limits_{S} \underline{\hspace{2cm}} \ d\sigma$$

$$= \iint\limits_{x^2+y^2 \leq 4} \underline{\hspace{2cm}}$$

$$= \underline{\hspace{2cm}} \ .$$

15.7 PATH INDEPENDENCE, POTENTIAL FUNCTIONS, AND CONSERVATIVE FIELDS

OBJECTIVE A : Know the meaning of a conservative field $\vec{F}$ throughout an open connected region D in space, and the basic implications of conservative fields.

44. A vector field $\vec{F}$ is said to be <u>conservative</u> on D if there is a scalar function f on D such that

$$\vec{F} = \underline{\hspace{2cm}} \ .$$

The function f is called a $\underline{\hspace{4cm}}$ for $\vec{F}$.

45. If $\vec{F}$ is conservative on D, then

$$\text{curl } \vec{F} = \underline{\hspace{3cm}}$$

throughout D.

46. Whenever $\vec{F}$ is conservative on D, the work integral over <u>any</u> closed path in D has value $\underline{\hspace{1.5cm}}$.

47. If the components of $\vec{F} = M\vec{i} + N\vec{j} + P\vec{k}$ are continuous throughout D, then $\vec{F}$ is conservative if and only if for all points A and B in D the value of

$$\int_A^B \vec{F} \cdot d\vec{r}$$

is $\underline{\hspace{4cm}}$ joining A to B in D.

---

43. $x^2 + y^2 + z^2 = 4$, $\vec{i} + \vec{j} - \vec{k}$, $\dfrac{x\vec{i} + y\vec{j} + z\vec{k}}{2}$, $-\dfrac{z}{2}$, $- dA$, -4π

44. $\vec{\nabla}f$, potential function   45. the zero vector $\vec{0}$ 46. zero

47. independent of the path

48. If the integral is independent of the path from A to B, its value is

$$\int_A^B \vec{F} \cdot d\vec{r} = \underline{\hspace{4cm}}$$

where $\vec{F} = \underline{\hspace{2cm}}$.

49. Whenever a field $\vec{F}$ is conservative, the integrand in the work integral,

$$\vec{F} \cdot d\vec{r} = M dx + N dy + P dz$$

is an $\underline{\hspace{5cm}}$.

50. If $M(x,y,z)$, $N(x,y,z)$, and $P(x,y,z)$ are continuous, together with their $\underline{\hspace{3cm}}$ partial derivatives, then a necessary and sufficient condition for the expression

$$M dx + N dy + P dz$$

to be an exact differential is that the following equations all be satisfied:

$$\underline{\hspace{2cm}}, \quad \underline{\hspace{2cm}}, \quad \underline{\hspace{2cm}} .$$

$\boxed{\text{OBJECTIVE B}}$: Determine if a specified force field $\vec{F} = M\vec{i} + N\vec{j} + P\vec{k}$ is conservative. If so, find a potential function f for the field.

51. Consider the vector field

$$\vec{F}(x,y,z) = (2xyz)\vec{i} + (x^2 z)\vec{j} + (x^2 y + 1)\vec{k} .$$

Let us determine if $\vec{F}$ is conservative. First,
$\vec{F} \cdot d\vec{r} = (2xyz)dx + (x^2 z)dy + (x^2 y + 1)dz$. Then, applying the test in Problem 50 above, with $M = 2xyz$, $N = \underline{\hspace{1.5cm}}$, and $P = x^2 y + 1$, we obtain

$$\frac{\partial M}{\partial z} = 2xy = \frac{\partial P}{\partial x}, \quad \frac{\partial M}{\partial y} = \underline{\hspace{1.5cm}} = \frac{\partial N}{\partial x}, \quad \text{and} \quad \frac{\partial}{\partial y}\underline{\hspace{0.8cm}} = x^2 = \frac{\partial N}{\partial z} .$$

Therefore, we conclude that $\vec{F}$ is conservative and there is a function $f(x,y,z)$ such that $\vec{F} \cdot d\vec{r} = df$. We want to find the potential function f. Now f satisfies

$$\frac{\partial f}{\partial x} = 2xyz, \quad \frac{\partial f}{\partial y} = \underline{\hspace{1.5cm}}, \quad \text{and} \quad \frac{\partial f}{\partial z} = \underline{\hspace{2cm}} .$$

To find f we integrate the first of these equations with respect to x, holding y and z constant, obtaining

---

48. f(B) - f(A), $\vec{F} = \vec{\nabla} f$ 49.   exact differential

50. first-order, $\frac{\partial M}{\partial y} = \frac{\partial N}{\partial x}, \quad \frac{\partial M}{\partial z} = \frac{\partial P}{\partial x}, \quad \frac{\partial N}{\partial z} = \frac{\partial P}{\partial y}$

$$f(x,y,z) = \underline{\hspace{2cm}} + g(y,z) \ ,$$

where $g(y,z)$ is an arbitrary function acting as the "constant of integration." Next, we differentiate the last equation with respect to y, holding x and z constant, equate this to $\frac{\partial f}{\partial y}$, and solve for $\frac{\partial g}{\partial y}$:

$$\underline{\hspace{2cm}} + \frac{\partial g(y,z)}{\partial y} = x^2 z, \quad \text{so that} \quad \frac{\partial g(y,z)}{\partial y} = \underline{\hspace{1.5cm}} \ .$$

Therefore, $g(y,z) = h(z)$ is a function of z alone. The expression for f then becomes

$$f(x,y,z) = \underline{\hspace{2cm}} + h(z) \ .$$

We differentiate this last equation with respect to z, holding x and y constant, equate this to $\frac{\partial f}{\partial z}$, and solve for $h'(z)$:

$$\underline{\hspace{2cm}} + h'(z) = x^2 y + 1, \quad \text{so that} \quad h'(z) = \underline{\hspace{1.5cm}} \ .$$

Therefore, $h(z) = \underline{\hspace{1.5cm}} + c$ for some arbitrary constant c. Hence we may write f as

$$f(x,y,z) = \underline{\hspace{2.5cm}} \ .$$

It is easy to verify that $\vec{F} = \vec{\nabla} f$.

52. Suppose it is required to find the line integral $\displaystyle\int_C \vec{F} \cdot d\vec{r}$ for the force field $\vec{F}$ in Problem 51 above, where C is composed of the line segments $A(0,0,0)$ to $(0,-1,-3)$ and from $(0,-1,-3)$ to $B(-1,1,2)$. Then,

$$\int_C \vec{F} \cdot d\vec{r} = f(\underline{\hspace{1.5cm}}) - f(\underline{\hspace{2cm}}) = \underline{\hspace{1.5cm}} \ .$$

---

51. x^2z, $2xz$, P, x^2z, $x^2y + 1$, x^2yz, x^2z, 0, x^2yz, x^2y, 1, z, $x^2yz + z + c$

52. $(-1,1,2)$, $(0,0,0)$, 4

CHAPTER 15 SELF-TEST

1. Evaluate the following line integrals:

   (a) $\int_C \sqrt{x + z}\; ds$, where C is the line segment from the point $A(0,0,0)$ to $B(4,5,6)$

   (b) $\int_C y^2 dx + x^2 dy$, where C is the right semi-circle $x = \sqrt{1 - y^2}$ directed from the point $A(0,-1)$ to $B(0,1)$

2. Find the work done by the force

   $$\vec{F} = (\frac{1}{x + 3y + 2z} - 5x)\vec{i} + (\frac{3}{x + 3y + 2z} + 3y^2)\vec{j} + (\frac{2}{x + 3y + 2z})\vec{k}$$

   as the point of application moves from $A(0,2,0)$ along the y-axis to $(0,1,0)$, and from there in a straight line to $B(1,1,1)$. <u>Hint</u>: Is the field conservative?

3. Use Green's theorem to evaluate the line integral $\oint_C y\, dx - x\, dy$,

   where C is the cardioid $r = 1 - \cos\theta$ directed in counter-clockwise direction.

4. Use Green's theorem to find the area of the region bounded by the hypocycloid parameterized by $x = a\cos^3 t$, $y = a\sin^3 t$, for $0 \leq t \leq 2\pi$. Assume the boundary is directed in the counterclock-wise direction.

5. A force is given by $\vec{F} = (x^2 - xy^3)\vec{i} + (y^2 - 2xy)\vec{j}$, and its point of application moves counterclockwise around the square in the xy-plane with vertices $(0,0)$, $(3,0)$, $(3,3)$, and $(0,3)$. Find the work done.

6. Find the surface area of that part of the cylinder $x^2 + z^2 = a^2$ that lies inside the cylinder $x^2 + y^2 = a^2$.

7. Let S be the portion of the cylinder $z = \frac{1}{2}y^2$ for which $x \geq 0$, $y \geq 0$, and $x + y \leq 1$. Find the surface integral

   $$\iint_S \sqrt{x(1 + 2z)}\; d\sigma\; .$$

8. Evaluate the flux integral

   $$\iint_S \vec{F} \cdot \vec{n}\; d\sigma\; ,$$

   for $\vec{F} = x\vec{i} + y\vec{j} + z\vec{k}$, given that S is the plane of the triangle with vertices $(a,0,0)$, $(0,a,0)$, $(0,0,a)$, $a > 0$, lying in the first octant, and $\vec{n}$ is the upper normal unit vector.

9. Use the divergence theorem to find the flux of $\vec{F}$ outward through the region inside the sphere $x^2 + y^2 + z^2 = 4$ and outside the cylinder $x^2 + y^2 = 1$, if the vector flow field is $\vec{F} = 6x\vec{i} - 13y\vec{j} + 12z\vec{k}$.

10. Let C be the intersection of the sphere $x^2 + y^2 + z^2 = 1$ and the cone $z = \sqrt{x^2 + y^2}$, directed in the counterclockwise sense around the z-axis. For the field
$$\vec{F} = (x^2 + z)\vec{i} + (y^2 + 2x)\vec{j} + (z^2 - y)\vec{k} ,$$
use Stokes's theorem to calculate the circulation of $\vec{F}$ around C.

11. Find the flux of the vector field $\vec{F} = -y\vec{i} + x\vec{j} + z^4\vec{k}$ outward through the sphere $x^2 + y^2 + z^2 = 4$.

12. Determine if the force field
$$\vec{F} = (2xz + 1)\vec{i} + 2y(z + 1)\vec{j} + (x^2 + y^2 + 3z^2)\vec{k}$$
is conservative. If so, find a potential function f for the field.

SOLUTIONS TO CHAPTER 15 SELF-TEST

1. (a) A parametric representation for the line segment   C   is
        $x = 4t$, $y = 5t$,   and $z = 6t$ with $0 \le t \le 1$.   Thus,

$$\int_C \sqrt{x + z}\ ds = \int_0^1 \sqrt{4t + 6t} \cdot \sqrt{\left(\frac{dx}{dt}\right)^2 + \left(\frac{dy}{dt}\right)^2 + \left(\frac{dz}{dt}\right)^2}\ dt$$

$$= \int_0^1 \sqrt{10t}\ \sqrt{77}\ dt = \frac{2}{3}\sqrt{770} \approx 18.50.$$

   (b) A parametric representation for the semi-circle is
        $x = \cos\theta$ and $y = \sin\theta$ with $-\frac{\pi}{2} \le \theta \le \frac{\pi}{2}$.   (Note the
        orientation.)   Thus,

$$\int_C y^2\,dx + x^2\,dy = \int_{-\pi/2}^{\pi/2} [\sin^2\theta(-\sin\theta) + \cos^2\theta(\cos\theta)]\ d\theta$$

$$= \int_{-\pi/2}^{\pi/2} (\cos^3\theta - \sin^3\theta)\ d\theta = \int_{-\pi/2}^{\pi/2} \cos^3\theta\ d\theta$$

$$\uparrow$$

(since $\sin^3\theta$ is an odd function)

$$= \tfrac{1}{3}(\sin\theta)(\cos^2\theta + 2)]_{-\pi/2}^{\pi/2} = \tfrac{4}{3}\ .$$

2. From a little observation of the components of $\vec{F}$,   it is easy to
    see that $\vec{F} = \vec{\nabla}f$,   where

$$f(x,y,z) = \ln|x + 3y + 2z| - \tfrac{5}{2}x^2 + y^3\ .$$

Therefore, $\vec{F}$ is a conservative force field so that the work
done by $\vec{F}$ from   A   to   B   is independent of the path joining
them.   The work done is

$$W = \int_A^B \vec{F} \cdot d\vec{r} = f(1,1,1) - f(0,2,0)$$

$$= (\ln 6 - \tfrac{5}{2} + 1) - (\ln 6 - 0 + 8) = -\tfrac{19}{2}\ .$$

3. Let   R   denote the region that lies inside the cardioid.   Then,

$$\oint_C y\,dx - x\,dy = -2 \iint_R dx\ dy = -2 \int_0^{2\pi} \int_0^{1-\cos\theta} r\ dr\ d\theta$$

L note the orientation and
polar coordinates

$$= \int_{2\pi}^0 (1 - 2\cos\theta + \cos^2\theta)\ d\theta$$

$$= \theta - 2\sin\theta + \tfrac{\theta}{2} + \tfrac{1}{4}\sin 2\theta]_{2\pi}^0 = -3\pi\ .$$

4. A sketch of the region is given at
    the right. From the symmetry of
    the region we can calculate the
    area of the region in the first
    quadrant and multiply by 4. We
    use the area formula from Green's
    theorem. For the parameterization

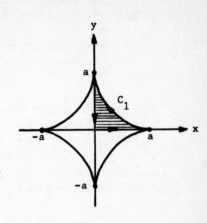

$x = a \cos^3 t$ and $y = a \sin^3 t$,

$dx = -3a \sin t \cos^2 t \, dt$ and

$dy = 3a \cos t \sin^2 t \, dt$.

Thus, the area of the entire region
is given by

$$A = \oint \tfrac{1}{2} (x \, dy - y \, dx)$$

$$= 4 \int_{C_1} \tfrac{1}{2}(x \, dy - y \, dx)$$

$$= 4 \int_0^{\pi/2} \tfrac{1}{2}(3a^2 \cos^4 t \sin^2 t + 3a^2 \sin^4 t \cos^2 t) \, dt$$

$$= 6a^2 \int_0^{\pi/2} \cos^2 t \sin^2 t \, dt = 6a^2 [\tfrac{t}{8} - \tfrac{1}{32} \sin 4t]_0^{\pi/2} = \tfrac{3\pi}{8} a^2 \; .$$

$\uparrow$_ (as in Problem 14 of this chapter)

5. According to Green's theorem, the work done is

$$W = \oint_C \vec{F} \cdot d\vec{r} = \int_0^3 \int_0^3 [\tfrac{\partial}{\partial x}(y^2 - 2xy) - \tfrac{\partial}{\partial y}(x^2 - xy^3)] \, dy \, dx$$

$$= \int_0^3 \int_0^3 (-2y + 3xy^2) \, dy \, dx$$

$$= \int_0^3 (-9 + 27x) \, dx = \tfrac{189}{2} \; .$$

6. A sketch of the intersecting cylinders is below. By the
    symmetry of the surface we need only calculate the area of the
    top surface (above the xy-plane), and multiply this by 2. In
    that case, $z = \sqrt{a^2 - x^2}$ and this yields

$\dfrac{\partial z}{\partial x} = \dfrac{-x}{\sqrt{a^2 - x^2}}$ and $\dfrac{\partial z}{\partial y} = 0$. Therefore,

$$\sqrt{(f_x)^2 + (f_y)^2 + 1}$$

$$= \sqrt{\frac{x^2}{a^2 - x^2} + 1} = \frac{a}{\sqrt{a^2 - x^2}} .$$

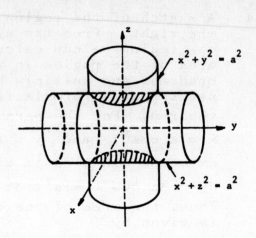

The region in the xy-plane over which the surface lies is the interior of the circle $x^2 + y^2 = a^2$. Thus, the total surface area is given by the double integral,

$$S = 2 \int_{-a}^{a} \int_{-\sqrt{a^2-x^2}}^{\sqrt{a^2-x^2}} \frac{a}{\sqrt{a^2 - x^2}} \, dy \, dx$$

$$= 2a \int_{-a}^{a} 2 \, dx = 8a^2 .$$

7.   The portion of the cylindrical surface is sketched below. For the element of surface area $d\sigma$ we use

$$d\sigma = \sqrt{1 + \left(\frac{\partial z}{\partial x}\right)^2 + \left(\frac{\partial z}{\partial y}\right)^2} \, dx \, dy = \sqrt{1 + 0 + y^2} \, dx \, dy .$$

The surface integral becomes

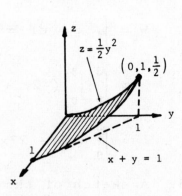

$$\iint_S \sqrt{x(1 + 2z)} \, d\sigma$$

$$= \int_0^1 \int_0^{1-x} \sqrt{x}(1 + y^2) \, dy \, dx$$

$$= \int_0^1 \sqrt{x}\left[(1 - x) + \frac{1}{3}(1 - x)^3\right] \, dx$$

$$= \int_0^1 \left(\frac{4}{3}x^{1/2} - 2x^{3/2} + x^{5/2} - \frac{1}{3}x^{7/2}\right) \, dx$$

$$= \frac{8}{9} - \frac{4}{5} + \frac{2}{7} - \frac{2}{27} \approx 0.30 .$$

8. The triangular shaped surface S is shown at the right. An equation of the plane is

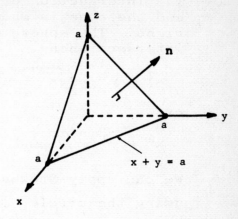

$$x + y + z = a, \quad \text{or} \quad z = a - x - y.$$

A vector normal to the plane is $\vec{N} = \vec{i} + \vec{j} + \vec{k}$, and this vector points in the upward direction. Thus,

$$\vec{n} = \frac{\vec{i} + \vec{j} + \vec{k}}{\sqrt{3}}.$$

For the element of surface area we have,

$$d\sigma = \sqrt{1 + \left(\frac{\partial z}{\partial x}\right)^2 + \left(\frac{\partial z}{\partial y}\right)^2}$$

$$= \sqrt{3} \; dx \; dy.$$

The projection of the surface onto the xy-plane is the region described by $x \geq 0$, $y \geq 0$, and $x + y \leq a$ (see figure). Thus,

$$\iint\limits_{S} \vec{F} \cdot \vec{n} \; d\sigma = \iint\limits_{S} \frac{1}{\sqrt{3}}(x + y + z) \; d\sigma$$

$$= \int_0^a \int_0^{a-x} (x + y + a - x - y) \; dy \; dx$$

$$= a \int_0^a (a - x) \; dx = \tfrac{1}{2}a^3.$$

9. Evaluation of $\iint\limits_{S} \vec{F} \cdot \vec{n} \; d\sigma$ itself would involve finding two surface integrals, one for the sphere and one for the inside cylinder. Instead we use the divergence theorem:

$$\iint\limits_{S} \vec{F} \cdot \vec{n} \; d\sigma = \iiint\limits_{D} \text{div} \; \vec{F} \; dV = \iiint\limits_{D} (6 - 13 + 12) \; dV$$

cylindrical ↑
coordinates

$$= 5 \int_0^{2\pi} \int_1^2 \int_{-\sqrt{4-r^2}}^{\sqrt{4-r^2}} dz \; r \; dr \; d\theta$$

$$= 5 \int_0^{2\pi} \int_1^2 2r\sqrt{4 - r^2} \; dr \; d\theta$$

$$= 5 \int_0^{2\pi} -\tfrac{2}{3}(4 - r^2)^{3/2} \big]_{r=1}^{r=2} \; d\theta = 20\pi\sqrt{3} \approx 108.83.$$

10. The intersection of the sphere
     and the cone is shown at the
     right. The sphere and cone
     intersect when

$$x^2 + y^2 + \left(\sqrt{x^2 + y^2}\right)^2 = 1$$

or

$$x^2 + y^2 = \tfrac{1}{2} \quad \text{and} \quad z = \tfrac{1}{\sqrt{2}} \ .$$

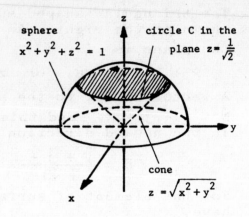

sphere
$x^2 + y^2 + z^2 = 1$

circle C in the
plane $z = \tfrac{1}{\sqrt{2}}$

cone
$z = \sqrt{x^2 + y^2}$

We can apply Stokes's theorem
using the circle C: $x^2 + y^2 = \tfrac{1}{2}$
considered in the <u>plane</u> $z = \tfrac{1}{\sqrt{2}}$
(rather than using the
spherical cap above it for S.)
The reason for this choice of the
surface is that its unit normal
is the constant vector $\vec{n} = \vec{k}$ (rather than the varying unit
normal on the cap). Then,

$$\text{curl } \vec{F} = \begin{vmatrix} \vec{i} & \vec{j} & \vec{k} \\ \dfrac{\partial}{\partial x} & \dfrac{\partial}{\partial y} & \dfrac{\partial}{\partial z} \\ x^2+z & y^2+2x & z^2-y \end{vmatrix} = -\vec{i} + \vec{j} + 2\vec{k} \ ,$$

and Stokes's theorem gives the circulation of $\vec{F}$ around C as

$$\oint_C \vec{F} \cdot d\vec{r} = \iint_S \text{curl } \vec{F} \cdot \vec{n} \ d\sigma$$

$$= \iint_S 2 \ d\sigma = 2 \cdot \text{area enclosed by the circle C}$$

$$= 2 \cdot \pi \left(\tfrac{1}{\sqrt{2}}\right)^2 = \pi \ .$$

11. We divide the sphere into the upper hemisphere

S_1: $z = \sqrt{4 - x^2 - y^2}$ and the lower hemisphere

S_2: $z = -\sqrt{4 - x^2 - y^2}$. In either case, a unit outer normal is
given by

$$\vec{n} = \frac{x\vec{i} + y\vec{j} + z\vec{k}}{\sqrt{x^2 + y^2 + z^2}} = \frac{\vec{\nabla}w}{|\vec{\nabla}w|} \quad \text{for} \quad w = x^2 + y^2 + z^2 - 4 \ .$$

Thus, $\vec{n} = \tfrac{1}{2}(x\vec{i} + y\vec{j} + z\vec{k})$ since $x^2 + y^2 + z^2 = 4$ on the
sphere. Notice that $\vec{n}$ points in the outward direction
(upward for $z > 0$ and downward for $z < 0$).

An element of surface area for S_1 or S_2 is given by

$$d\sigma = \sqrt{1 + \left(\frac{\partial z}{\partial x}\right)^2 + \left(\frac{\partial z}{\partial y}\right)^2}\ dx\ dy = \sqrt{1 + \frac{x^2}{z^2} + \frac{y^2}{z^2}}\ dx\ dy$$

$$= \sqrt{\frac{x^2 + y^2 + z^2}{z^2}}\ dx\ dy = \frac{2}{|z|}\ .$$

Thus, the flux of $\vec{F}$ outward through the sphere S is,

$$\iint\limits_{S_1 \cup S_2} \vec{F} \cdot \vec{n}\ d\sigma = \iint\limits_{S_1} \vec{F} \cdot \vec{n}\ d\sigma + \iint\limits_{S_2} \vec{F} \cdot \vec{n}\ d\sigma$$

$$= \iint\limits_{S_1} \tfrac{1}{2} z^5 \cdot \frac{2}{|z|}\ dx\ dy + \iint\limits_{S_2} \tfrac{1}{2} z^5 \cdot \frac{2}{|z|}\ dx\ dy$$

$$= \iint\limits_{x^2 + y^2 \le 4} z^4\ dx\ dy + \iint\limits_{x^2 + y^2 \le 4} - z^4\ dx\ dy = 0\ .$$

12. If $M = 2xz + 1$, $N = 2y(z + 1)$, and $P = x^2 + y^2 + 3z^2$, then

$$\frac{\partial M}{\partial z} = 2z = \frac{\partial P}{\partial x}, \quad \frac{\partial N}{\partial z} = 2y = \frac{\partial P}{\partial y}, \quad \frac{\partial M}{\partial y} = 0 = \frac{\partial N}{\partial x}\ .$$

Thus, the field is conservative and there is a potential function $f(x,y,z)$ with $\vec{F} = \vec{\nabla} f$. Now, f satisfies

$$\frac{\partial f}{\partial x} = 2xz + 1, \quad \frac{\partial f}{\partial y} = 2y(z + 1), \quad \text{and} \quad \frac{\partial f}{\partial z} = x^2 + y^2 + 3z^2\ .$$

Integration of the first of these equations with respect to x, holding y and z fixed, gives

"constant of integration"
$\downarrow$

$$f(x,y,z) = x^2 z + x + g(y,z)\ .$$

Then,

$$2y(z + 1) = \frac{\partial f}{\partial y} = \frac{\partial g(y,z)}{\partial y} \quad \text{so} \quad g(y,z) = y^2(z + 1) + h(z)\ ,$$

where $h(z)$ is the "constant of integration" obtained when $\frac{\partial g}{\partial y} = 2y(z + 1)$ is integrated with respect to y with z fixed. Then,

$$f(x,y,z) = x^2 z + x + y^2(z + 1) + h(z)\ ,$$

and

$$x^2 + y^2 + 3z^2 = \frac{\partial f}{\partial z} = x^2 + y^2 + h'(z)$$

so $h'(z) = 3z^2$. Hence, $h(z) = z^3 + c$ and

$$f(x,y,z) = x(xz + 1) + y^2(z + 1) + z^3 + c,$$

where c is an arbitrary constant.

NOTES:

CHAPTER 16 PREVIEW OF DIFFERENTIAL EQUATIONS

16.1 SEPARABLE FIRST ORDER EQUATIONS

1. A differential equation is an equation that contains one or more _____, or _____ .

2. Differential equations are classified by _____ and _____ .

3. The <u>order</u> of a differential equation is the order of the highest _____ that occurs in the equation.

4. A differential equation is _____ if it can be put in the form

$$a_n(x)\frac{d^n y}{dx^n} + a_{n-1}(x)\frac{d^{n-1}y}{dx^{n-1}} + \cdots + a_1(x)\frac{dy}{dx} + a_0(x)y = F(x)$$

    where the a's are functions of x.

5. The differential equation

$$x\frac{d^3 y}{dx^3} + \left(\frac{dy}{dx}\right)^2 - e^x y = 0$$

    is an _____ differential equation of order _____ . It is not _____ .

6. The differential equation

$$\frac{\partial u}{\partial t} = \left(h^2\frac{\partial^2 u}{\partial x^2} + \frac{\partial^2 u}{\partial y^2}\right)$$

    is a _____ differential equation of order _____ .

7. A function $y = f(x)$ is said to be a _____ of a differential equation if the latter is satisfied when _____ and its _____ are replaced throughout by _____ and its corresponding derivatives.

OBJECTIVE A : Show that a given function is a solution to a specified (ordinary) differential equation.

8. Consider the differential equation
$$3xy' - y = \ln x + 1, \quad x > 0 ,$$

---

1. derivatives, differentials   2. type, order   3. derivative   4. linear

5. ordinary, three, linear   6. partial, two   7. solution, y, derivatives, f(x)

and the function

$$y = f(x) = Cx^{1/3} - \ln x - 4,$$

where   C   is any constant.   Then,

$$\frac{dy}{dx} = f'(x) = \underline{\hspace{3cm}},$$

and

$$3xy' = \underline{\hspace{3cm}} .$$

Thus,

$$3xy' - y = (Cx^{1/3} - 3) - (Cx^{1/3} - \ln x - 4)$$
$$= \underline{\hspace{2cm}} .$$

Therefore, the function $f(x) = Cx^{1/3} - \ln x - 4$, and its derivative, satisfy the differential equation.   We have verified that $y = f(x)$ is a solution.

9.   Consider the second order differential equation

$$2xy'' + (1 - 4x)y' + (2x - 1)y = e^x ,$$

and the function

$$y = f(x) = (c_1 + c_2\sqrt{x} + x)e^x, \quad x > 0,$$

where c_1 and c_2 are any constants.   Then,

$$y' = (\underline{\hspace{4cm}}) e^x ,$$

and

$$y'' = (\underline{\hspace{4cm}}) e^x .$$

Then,

$$2xy'' = [(2c_1 + 4)x + 2x^2 - \tfrac{1}{2}c_2x^{-1/2} + 2c_2x^{1/2} + 2c_2x^{3/2}]e^x ,$$

$$(1 - 4x)y' = [\underline{\hspace{7cm}}]e^x ,$$

$$(2x - 1)y = [-c_1 + (2c_1 - 1)x + 2x^2 - c_2x^{1/2} + 2c_2x^{3/2}]e^x .$$

Therefore,

$$2xy'' + (1 - 4x)y' + (2x - 1)y = \underline{\hspace{2cm}} .$$

We conclude that $y = (c_1 + c_2\sqrt{x} + x)e^x$ _____ a solution to the differential equation.

---

8. $\frac{1}{3}Cx^{-2/3} - \frac{1}{x}$, $Cx^{1/3} - 3$, $\ln x + 1$

9. $c_1 + c_2\sqrt{x} + x + 1 + \frac{1}{2}c_2x^{-1/2}$, $c_1 + c_2\sqrt{x} + x + 2 + c_2x^{-1/2} - \frac{1}{4}c_2x^{-3/2}$,

   $c_1 + 1 - (4c_1 + 3)x - 4x^2 + \frac{1}{2}c_2x^{-1/2} - c_2x^{1/2} - 4c_2x^{3/2}$, e^x,   is

OBJECTIVE B : Solve first order differential equations in which the variables can be separated. If initial conditions are prescribed, determine the value of the constant of integration.

10. Solve the differential equation

$$(xy - x)\ dx + (xy + y)\ dy = 0\ .$$

Solution. We separate the variables and integrate:

$$x(y - 1)\ dx + y(x + 1)\ dy = 0\ ,$$

$$\frac{y\ dy}{y - 1} = \underline{\hspace{3cm}}\ .$$

Then,

$$\left(1 + \frac{1}{y - 1}\right)\ dy = \underline{\hspace{3cm}}$$

$$y + \ln|y - 1| = \underline{\hspace{2.5cm}} + \ln\ C.$$

We introduce $\ln C$, $C > 0$, as the constant of integration in order to simplify the form of the solution. Thus, by algebra,

$$x + y = \ln\ C\left|\frac{x + 1}{y - 1}\right| \quad or, \quad |y - 1|e^{x+y} = \underline{\hspace{2cm}}\ .$$

11. Solve the differential equation

$$x^2 yy' = e^y\ ; \quad when \quad x = 2, \quad y = 0\ .$$

Solution: We change to differential form, separate the variables, and integrate:

$$x^2 y\ dy = e^y\ dx \quad or, \quad ye^{-y}dy = \underline{\hspace{2.5cm}}\ .$$

We integrate the left side by parts, using $u = y$ and $dv = e^{-y}dy$:

$$\underline{\hspace{4cm}} = -\frac{1}{x} + C$$

$$-e^{-y}(\underline{\hspace{2cm}}) = -\frac{1}{x} + C\ ,$$

or simplifying algebraically,

$$x(y + 1) = \underline{\hspace{2.5cm}}\ .$$

Using the initial condition $x = 2$ and $y = 0$ gives,

$$2(0 + 1) = \underline{\hspace{2cm}}, \quad or \quad C = \underline{\hspace{2cm}}\ .$$

Thus, the solution is given by $x(y + 1) = \left(1 + \frac{x}{2}\right)e^y$.

---

10. $-\frac{x\ dx}{x + 1}$, $-\left(1 - \frac{1}{x + 1}\right)\ dx$, $-x + \ln|x + 1|$, $C|x + 1|$

11. $x^{-2}\ dx$, $-ye^{-y} + \int e^{-y}dy$, $y + 1$, $(1 - Cx)e^y$, $1 - 2C$, $-\frac{1}{2}$

[OBJECTIVE C]:   Determine if a differential equation is homogeneous, and if it is, solve it.

12.   If a differential equation can be put into the form
$\frac{dy}{dx}$ = _____ ,   then the equation is called <u>homogeneous</u>. The
equation becomes separable in the variables x and v by
defining v = _____ .

13.   Solve the differential equation

$$3xy^2 \, dy = (4y^3 - x^3) \, dx \ .$$

<u>Solution</u>.   From the given equation, we have

$$\frac{dy}{dx} = \frac{4y^3 - x^3}{3xy^2} = \tfrac{1}{3}[4\left(\tfrac{y}{x}\right) - \underline{\hspace{1.5cm}}] \ .$$

The equation is homogeneous, and we let $v = \tfrac{y}{x}$,   or $y = vx$.
Then $\frac{dy}{dx}$ = _____ and the differential equation becomes,

$$v + x \frac{dv}{dx} = \underline{\hspace{3cm}} \ ,$$

or separating the variables x and v,

$$\frac{dx}{x} = 3\left(\underline{\tfrac{v^2 \, dv}{\hspace{1.5cm}}}\right) \ .$$

The solution of this is,

$$\ln |x| = \underline{\hspace{4cm}} + \ln C$$

so that

$$|x| = C|v^3 - 1| \ .$$

In terms of x and y,   the solution is

$$\underline{\hspace{6cm}} \ .$$

[OBJECTIVE D]:   Find the family of solutions of a given differential
equation and the family of <u>orthogonal</u> <u>trajectories</u>
(defined on page 1011 in the Exercise section).

14.   Consider the differential equation $x + y \frac{dy}{dx} = 0$.   Separating
the variables, and integrating, gives

x dx + y dy = 0,   and _____ = C_1,   where $C_1 > 0$.

Thus, the family of solutions is a family of _____
centered at _____ .   To find the family of orthogonal
trajectories, we solve the differential equation _____ .

---

12. $F\left(\tfrac{y}{x}\right)$, $\tfrac{y}{x}$

13. $\left(\tfrac{x}{y}\right)^2$, $v + x\frac{dv}{dx}$, $\tfrac{1}{3}(4v - v^{-2})$, $v^3 - 1$, $\ln |v^3 - 1|$, $|x| = C\left|\tfrac{y^3}{x^3} - 1\right|$

Separating the variables, and integrating, gives

$$\frac{dy}{y} = \underline{\hspace{2cm}} \quad \text{and} \quad \ln|y| = \underline{\hspace{2cm}} + \ln C_2 ,$$

or

$$y = \pm C_2 x, \quad C_2 > 0 .$$

Notice that $y = 0$, the x-axis, also solves the differential equation for the orthogonal trajectories; so does $x = 0$, the y-axis. Thus, the orthogonal trajectories are the family of _____ passing through the origin. A sketch of the solution family, and the family of orthogonal trajectories, is given at the right.

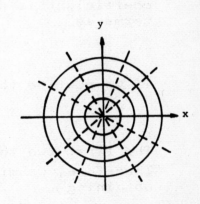

16.2 EXACT DIFFERENTIAL EQUATIONS

OBJECTIVE : Solve a differential equation that is exact. It might be necessary to make the given equation exact by multiplication by a suitable integrating factor $\rho(x,y)$.

15. Show the differential equation

$$\cos y \, dx - (x \sin y - y^2) \, dy = 0$$

is exact and solve it.

Solution. $\frac{\partial}{\partial y}(\cos y) = -\sin y$ and $\frac{\partial}{\partial x}(-x \sin y + y^2) =$ _____ so the equation is exact: that is, the left side is an exact differential $df(x,y)$. Now, $\frac{\partial f}{\partial x} = $ _____ so $f(x,y) = $ _____ $+ g(y)$. Differentiating the last equation with respect to y with x held constant gives,

$$\frac{\partial f}{\partial y} = \underline{\hspace{4cm}} .$$

Thus, $g'(y) = $ _____ . Hence, $g(y) = $ _____ $+ C_1$ and

$$f(x,y) = -x \sin y + \frac{y^3}{3} + C_1 .$$

Therefore,

$$-x \sin y + \frac{y^3}{3} = C ,$$

where C is an arbitrary constant, solves the differential equation.

---

14. $x^2 + y^2$, circles, the origin, $y \, dx - x \, dy = 0$, $\frac{dx}{x}$, $\ln|x|$, lines

15. $-\sin y$, $\cos y$, $x \cos y$, $-x \sin y + g'(y)$, y^2, $\frac{1}{3} y^3$

16.   Let us solve $3x^2y \, dx + (y^4 - x^3) \, dy = 0$.

Since $\frac{\partial}{\partial y}(3x^2y) = 3x^2 \neq -3x^2 = \frac{\partial}{\partial x}(y^4 - x^3)$,   the equation, as it stands, is not exact. However, two terms in the coefficients of   dx   and   dy   are of degree three and the other coefficient is not of degree three, so we try regrouping the terms as

$$(3x^2y \, dx - x^3 \, dy) + y^4 \, dy = 0 \ .$$

If we rewrite this last equation as

$$[y \, d(x^3) - x^3 \, dy] + y^4 \, dy = 0 \ ,$$

The first two terms are suggestive of $d(\frac{u}{v}) = $ _____.
Therefore, we might divide the rewritten equation by   _____,
obtaining

$$\frac{y \, d(x^3) - x^3 dy}{y^2} + y^2 \, dy = 0.$$

Thus,

$$d(\underline{\hspace{1cm}}) + y^2 dy = 0 \quad \text{gives} \quad \underline{\hspace{2cm}} + C \ .$$

16.3   LINEAR FIRST ORDER EQUATIONS

OBJECTIVE A :   Determine if a differential equation of first order is linear, and if it is, solve it.

17.   A differential equation of first order, which is linear in the dependent variable   y,   can always be put in the standard form

_____ ,

where   P   and   Q   are functions of   x.

18.   Assuming that   P   and   Q   are continuous functions of   x,   we can solve a linear differential equation $y' + Py = Q$ by finding an <u>integrating</u> <u>factor</u>,

$$\rho(x) = \underline{\hspace{2cm}} \ ,$$

providing a solution $\rho y = $ _____.

---

16. $\dfrac{v \, du - u \, dv}{v^2}$, y^2, $\dfrac{x^3}{y}$, $\dfrac{x^3}{y} + \frac{1}{3} y^3$

17. $\dfrac{dy}{dx} + Py = Q$

18. $e^{\int P(x)dx}$, $\displaystyle\int \rho(x)Q(x) \, dx + C$

19. Let us solve the equation $x\dfrac{dy}{dx} + (x - 2)y = 3x^3 e^{-x}$. In standard form,

$$\frac{dy}{dx} + \left(1 - \frac{2}{x}\right)y = 3x^2 e^{-x} .$$

Here P = _____ and Q = _____, and the differential equation is linear. An integrating factor is given by

$$\rho = e^{\int P\,dx} = e^{\int (1 - \frac{2}{x})dx} = e^{\underline{\quad\quad}} = x^{\underline{\quad}} e^x .$$

Hence a solution is given by

$$x^{-2}e^x y = \int \underline{\hspace{2cm}} dx + C = \int \underline{\hspace{2cm}} dx + C$$

$$= \underline{\hspace{2cm}} .$$

Thus,

$$y = \underline{\hspace{3cm}} .$$

20. The differential equation $y' = x - 4xy$ can be written in standard form as

$$y' + 4xy = x ,$$

so it is linear. The equation may also be written in the form

$$\frac{dy}{1 - 4y} = \underline{\hspace{2cm}} ,$$

so it is separable in the variables x and y. Thus we have a choice of methods of solution. As a separable equation, we integrate the last equation, and find

$$-\frac{1}{4}\ln|1 - 4y| = \frac{1}{2}x^2 + \ln C ,$$

or

$$|1 - 4y| = C_1 \underline{\hspace{2cm}} , \quad\text{where}\quad C_1 = C^{-4} .$$

If we consider the differential equation as linear, an integrating factor is

$$\rho = e^{\int 4x\,dx} = \underline{\hspace{3cm}} ,$$

from which we get

$$ye^{2x^2} = \int \underline{\hspace{2cm}} dx + C_2 = \underline{\hspace{2cm}} + C_2$$

or

$$4y = \underline{\hspace{3cm}} .$$

---

19. $1 - \frac{2}{x}$, $3x^2 e^{-x}$, $x - 2\ln x$, -2, $x^{-2}e^x \cdot 3x^2 e^{-x}$, 3, $3x + C$, $(3x^3 + Cx^2)e^{-x}$

If $1 - 4y < 0$, we choose $4C_2 = C_1$, and if $1 - 4y \geq 0$, we choose $4C_2 = -C_1$. Thus both solution forms agree.

OBJECTIVE B : Solve second order differential equations in which the dependent variable is absent.

21. If a second order equation has the special form

$$F\left(x, \frac{dy}{dx}, \frac{d^2y}{dx^2}\right) = 0 ,$$

in which the _____ variable is missing, we can reduce it to a first order equation by substituting _____ and _____. (See Problems 15-19 on page 1019 of Finney/Thomas.)

22. To solve $\frac{d^2y}{dx^2} = x\left(\frac{dy}{dx}\right)^3$, we substitute $p = \frac{dy}{dx}$ and $\frac{dp}{dx} = \frac{d^2y}{dx^2}$. This gives, $\frac{dp}{dx} = $ _____ or $p^{-3}dp = $ _____, $p^{-2} = -x^2 + C_1{}^2$, $p = \frac{dy}{dx} = $ _____. Thus, $y + C_2 = \sin^{-1}\frac{x}{C_1}$ or $y + C_2 = $ _____. That is,

$$x = \underline{\qquad} \quad \text{or} \quad x = C_1 \cos(y + C_2) .$$

However, since $\cos(y + C_2) = \sin(y + C_2 + \frac{\pi}{2})$, and C_2 is an arbitrary constant, the second result is redundant, and we have $x = C_1 \sin(y + C_2)$ as the general solution.

16.4 SECOND ORDER LINEAR HOMOGENEOUS EQUATIONS

23. A <u>linear</u> equation of order n can be written in the form

_____ .

The coefficients $a_1, a_2, \ldots, a_n$ may be functions of x. If $F(x)$ is _____ the equation is said to be homogeneous; otherwise it is called _____.

24. The symbol D is introduced to represent differentiation with respect to _____, and powers of D mean taking successive derivatives:

$$D^2f(x) = \underline{\qquad} \quad \text{and} \quad D^nf(x) = \underline{\qquad}, \quad n \geq 2 .$$

---

20. $x\, dx$, e^{-2x^2}, e^{2x^2}, xe^{2x^2}, $\frac{1}{4}e^{2x^2}$, $1 + 4C_2e^{-2x^2}$ 21. dependent, $p = \frac{dy}{dx}$, $\frac{d^2y}{dx^2} = \frac{dp}{dx}$

22. xp^3, $x\, dx$, $\frac{\pm 1}{\sqrt{C_1{}^2 - x^2}}$, $\cos^{-1}\frac{x}{C_1}$, $C_1 \sin(y + C_2)$

23. $a_n\frac{d^ny}{dx^n} + a_{n-1}\frac{d^{n-1}y}{dx^{n-1}} + \cdots + a_1\frac{dy}{dx} + a_0y = F(x)$, identically zero, nonhomogeneous

24. x, $\frac{d^2f(x)}{dx^2}$, $\frac{d^nf(x)}{dx^n}$

25. Let us find $(2D^2 - D + 3)(e^x + \sin x)$:

$(2D^2 - D + 3)(e^x + \sin x)$

$= 2D^2(e^x + \sin x) - D(\underline{\hspace{2cm}}) + 3(e^x + \sin x)$

$= 2D(\underline{\hspace{1.5cm}}) - (e^x + \cos x) + 3(e^x + \sin x)$

$= 2(\underline{\hspace{1.5cm}}) + 2e^x + 3 \sin x - \cos x$

$= \underline{\hspace{4cm}}$.

26. Linear differential operators that are polynomials in D with constant coefficients satisfy basic algebraic laws that make it possible to treat them as $\underline{\hspace{3cm}}$ so far as $\underline{\hspace{1.5cm}}$, multiplication, and $\underline{\hspace{1.5cm}}$ are concerned.

$\boxed{\text{OBJECTIVE}}$:   Solve linear, second order, homogeneous equations with constant coefficients. The roots of the characteristic equation of the differential equation may be real and unequal, real and equal, or a pair of complex conjugate numbers.

27. Solve the equation

$$2\frac{d^2y}{dx^2} + 5\frac{dy}{dx} - 3y = 0 \ .$$

Solution. The associated characteristic equation is given by $\underline{\hspace{4cm}}$.   This equation factors into $(2r - 1)(\underline{\hspace{1.5cm}}) = 0$,   so the roots are $r_1 = \underline{\hspace{1.5cm}}$ and $r_2 = \underline{\hspace{1.5cm}}$.   Since these roots are real and unequal, the solution of the differential equation is

$$y = \underline{\hspace{5cm}} \ .$$

28. Solve the equation

$$\frac{d^2y}{dx^2} - 4\frac{dy}{dx} + 4y = 0 \ .$$

Solution. The characteristic equation is $\underline{\hspace{4cm}}$, and has roots $r_1 = \underline{\hspace{1.5cm}}$ and $r_2 = \underline{\hspace{1.5cm}}$.   Since these roots are real and equal, the solution of the differential equation is

$$y = \underline{\hspace{5cm}} \ .$$

---

25. $e^x + \sin x$, $e^x + \cos x$, $e^x - \sin x$, $4e^x + \sin x - \cos x$

26. ordinary polynomials, addition, factoring

27. $2r^2 + 5r - 3 = 0$, $r + 3$, $\frac{1}{2}$,   -3, $C_1e^{x/2} + C_2e^{-3x}$

28. $r^2 - 4r + 4 = 0$,   2,   2, $(C_1 + C_2x)e^{2x}$

29.   Solve the equation

$$\frac{d^2y}{dx^2} - 6\frac{dy}{dx} + 13y = 0 .$$

Solution. The characteristic equation is _____,
and has roots $r_1 =$ _____ and $r_2 =$ _____. Thus, these
roots are a pair of conjugate complex numbers with $\alpha =$ _____
and $\beta =$ _____. The solution of the differential equation is

$$y = \underline{\hspace{5cm}} .$$

16.5   SECOND ORDER NONHOMOGENEOUS LINEAR EQUATIONS WITH CONSTANT COEFFICIENTS

OBJECTIVE A :   Use the method of variation of parameters to solve
linear, second order, nonhomogeneous equations with
constant coefficients.

30.   A method for solving the nonhomogeneous equation

$$\frac{d^2y}{dx^2} + 2a\frac{dy}{dx} + by = F(x) ,$$

is first to obtain the general solution y_h of the related
homogeneous equation obtained by replacing _____.
Let this solution be denoted by

$$y_h = C_1 u_1(x) + C_2 u_2(x) .$$

Next, determine two functions $v_1 = v_1(x)$ and $v_2 = v_2(x)$ in
the following way: the derivatives v_1' and v_2' must satisfy
the two equations

_____ and _____ .

Solve this pair of simultaneous equations for v_1' and v_2', and
integrate these functions to obtain the functions $v_1 = v_1(x)$
and $v_2 = v_2(x)$ (don't forget the constants of integration).
Then, the general solution of the nonhomogeneous differential
equation is given by

$$y = \underline{\hspace{5cm}} .$$

The method described above is known as _____ .

---

29. $r^2 - 6r + 13 = 0,$ $3 + 2i,$ $3 - 2i,$ $3,$ $2,$ $e^{3x}(C_1 \cos 2x + C_2 \sin 2x)$

30. $F(x)$ by zero, $v_1'u_1 + v_2'u_2 = 0,$ $v_1'u_1' + v_2'u_2' = F(x),$ $v_1(x)u_1(x) + v_2(x)u_2(x),$

variation of parameters

31. Let us solve the equation $\dfrac{d^2y}{dx^2} - y = \dfrac{2}{e^x + 1}$ by variation of

parameters. We first solve the associated homogeneous

equation: $\dfrac{d^2y}{dx^2} - y = 0$ gives the characteristic equation

_____ , which has the roots $r_1 =$ _____ and

$r_2 =$ _____ . Thus, we find the solutions

$u_1(x) =$ _____ and $u_2(x) =$ _____

to the homogeneous equation. Next, we demand that the
functions v_1 and v_2 satisfy the equations

$$v_1' e^{-x} + v_2' e^x = 0, \quad \text{and} \quad \underline{\hspace{4cm}} .$$

Solving for v_1' and v_2' we find,

$$v_1' = \frac{\begin{vmatrix} 0 & e^x \\ \dfrac{2}{e^x+1} & e^x \end{vmatrix}}{\begin{vmatrix} e^{-x} & e^x \\ -e^{-x} & e^x \end{vmatrix}} = \frac{-e^x[2/(e^x + 1)]}{\underline{\hspace{2cm}}} = \underline{\hspace{3cm}} ,$$

and

$$v_2' = \frac{\begin{vmatrix} e^{-x} & 0 \\ -e^{-x} & \dfrac{2}{e^x+1} \end{vmatrix}}{\begin{vmatrix} e^{-x} & e^x \\ -e^{-x} & e^x \end{vmatrix}} = \frac{\underline{\hspace{3cm}}}{2} = \frac{e^{-x}}{e^x + 1} .$$

Integration then gives:

$$v_1 = \int \frac{-e^x \, dx}{e^x + 1} = \underline{\hspace{3cm}} + C_1 ,$$

and

$$v_2 = \int \frac{e^{-x} \, dx}{e^x + 1} = \int \frac{-u \, du}{u + 1} \qquad (u = e^{-x})$$

$$= \int \left(-1 + \frac{1}{u + 1}\right) du = \underline{\hspace{3cm}} + C_2 .$$

Therefore, the general solution is given by

$$y = v_1 u_1 + \underline{\hspace{3cm}}$$

$$= [-\ln(e^x + 1) + C_1]\underline{\hspace{1.5cm}} + [-e^{-x} + \ln(e^{-x} + 1) + C_2]\underline{\hspace{1.5cm}}$$

$$= C_1 e^{-x} + C_2 e^x - 1 - e^{-x} \ln(e^x + 1) + e^x \ln(e^{-x} + 1) .$$

---

31. $r^2 - 1 = 0$, -1, 1, e^{-x}, e^x, $-v_1' e^{-x} + v_2' e^x = \dfrac{2}{e^x + 1}$, 2, $\dfrac{-e^x}{e^x + 1}$, $\dfrac{2e^{-x}}{e^x + 1}$,

$-\ln(e^x + 1)$, $-e^{-x} + \ln(e^{-x} + 1)$, $v_2 u_2$, e^{-x}, e^x

OBJECTIVE B: Use the method of undetermined coefficients to solve linear, second order, nonhomogeneous equations with constant coefficients when the right-hand side is a sum of one or more terms like

$$e^{rx}, \quad \cos kx, \quad \sin kx, \quad ax^2 + bx + c .$$

32. Let's find a particular solution of $y'' + 3y' + 2y = 12x^2$. The characteristic equation is _____ and its roots are $r =$ _____ and $r =$ _____. Thus the general solution to the related homogeneous equation is $y_h =$ _____. To find a particular solution y_p that will produce $12x^2$ we choose our trial to be

$$y_p = Ax^2 + Bx + C$$

because _____ is not a root of the characteristic equation. Substitution of y_p, $y_p' = 2Ax + B$, and $y_p'' =$ _____ into the original equation yields

$$2A + 3(2Ax + B) + 2(Ax^2 + Bx + C) = 12x^2 .$$

This equation will hold if $2A =$ _____, $6A + 2B =$ _____, and _____ $= 0$. Thus, $A =$ _____, $B = -18$, and $C =$ _____. Our particular solution is _____.

16.6   OSCILLATION

OBJECTIVE: Find an equation for a given vibratory motion specified by a second order linear differential equation with constant coefficients. The motion may be damped or undamped.

33. An object weighing 2-lbs is suspended from the lower end of a spring, immersed in a medium, with its upper end attached to a rigid support. The object extends the spring 32 inches. After the object has come to rest in its equilibrium position, it is given an additional pull downward of 1 ft and released. As it moves up and down, its motion, taking into account the resistance of the medium, is described by the differential equation

$$m\frac{d^2x}{dt^2} = -kx - \frac{1}{2}\frac{dx}{dt} ,$$

where m is the mass of the object and k is the spring constant. Find its subsequent motion.

Solution. Let us first determine the spring constant k.

Since 2 lbs stretches the spring 32 in $= \frac{8}{3}$ ft, by Hooke's law $ks = mg$ we find,

---

32. $r^2 + 3r + 2 = 0$, -2, -1, $C_1e^{-2x} + C_2e^{-x}$, 0, $2A$, 12, 0, $2A + 3B + 2C$, 6, 21, $6x^2 - 18x + 21$

_____ = 2,   or   k = _____ .

The mass m of the object is m = _____,   and so the differential equation of the motion becomes

$$\frac{1}{16}\frac{d^2x}{dt^2} + \frac{1}{2}\frac{dx}{dt} + \frac{3}{4}x = 0 ,$$

or

$$\frac{d^2x}{dt^2} + 8\frac{dx}{dt} + 12x = 0 .$$

The characteristic equation is $r^2 + 8r + 12 = 0$,   and it has the two roots $r_1 = $ _____ and $r_2 = $ _____.   Thus, the general solution is

$$x = \text{_____} , \quad \text{with} \quad \frac{dx}{dt} = -2C_1e^{-2t} - 6C_2e^{-6t} .$$

The initial conditions are $t = 0$, $x = $ _____,   and $\frac{dx}{dt} = $ _____.   Substituting these values into the general solution and its derivative, we find

$$1 = C_1 + C_2 \quad \text{and} \quad \text{_____} .$$

Solving these equations simultaneously gives $C_1 = $ _____ and $C_2 = $ _____.   Therefore, an equation of the motion is

$$x = \frac{3}{2}e^{-2t} - \frac{1}{2}e^{-6t} .$$

Notice that as $t \rightarrow +\infty$, $x \rightarrow 0$.   The motion is nonoscillatory and we have _____ damping.

34.   For the same spring system as in Problem 33 above, suppose the resistance of the medium is such that the motion is described by the differential equation

$$m\frac{d^2x}{dt^2} = -kx - \frac{3}{8}\frac{dx}{dt} .$$

As before, $k = \frac{3}{4}$ and $m = \frac{1}{16}$ so the equation becomes

$$\frac{1}{16}\frac{d^2x}{dt^2} + \frac{3}{8}\frac{dx}{dt} + \frac{3}{4}x = 0 ,$$

or

$$\frac{d^2x}{dt^2} + 6\frac{dx}{dt} + 12x = 0 .$$

In this case the roots of the characteristic equation are $r_1 = $ _____ and $r_2 = $ _____.   Thus, the general solution is

$$x = \text{_____} ,$$

---

33. $\frac{8}{3}k$, $\frac{3}{4}$, $\frac{2}{32}$, -2, -6, $C_1e^{-2t} + C_2e^{-6t}$, 1, 0, $0 = -2C_1 - 6C_2$, $\frac{3}{2}$, $-\frac{1}{2}$, overcritical

or

$$x = Ce^{-3t} \sin(\sqrt{3}t + \phi) \ .$$

We will use the initial conditions $t = 0$, $x = 1$,   and $\frac{dx}{dt} = 0$ to determine the constants C and ϕ.

Differentiation of the general solution with respect to t yields

$$\frac{dx}{dt} = -3Ce^{-3t} \sin(\sqrt{3}t + \phi) + C\sqrt{3}e^{-3t} \cos(\sqrt{3}t + \phi) \ ,$$

and substitution of the values for the initial conditions into these equations for x and $\frac{dx}{dt}$ gives,

$$1 = \underline{\hspace{2cm}} \quad \text{and} \quad 0 = -3C \sin \phi + \sqrt{3}C \cos \phi \ .$$

Substituting $C = \csc \phi$ from the first of these equations into the second, we find

$$\cot \phi = \underline{\hspace{2cm}}, \quad \text{so} \quad \phi = \frac{\pi}{6} \ \text{or} \ \frac{7\pi}{6} \ .$$

Choosing $\phi = \frac{\pi}{6}$ and $\sin \phi = \frac{1}{2}$ gives $C = \underline{\hspace{2cm}}$. Therefore, the equation of the motion becomes

$$x = 2e^{-3t} \sin(\sqrt{3}t + \frac{\pi}{6}) \ .$$

As $t \to +\infty$, $x \to 0$. The motion is oscillatory and we have _____ damping. The damped period of the motion is $T = \underline{\hspace{2cm}}$.

16.7   NUMERICAL METHODS

35.   Consider the initial value problem $y' = f(x,y)$ and $y(x_0) = y_0$. Then Euler's method allows you to approximate the solution by stepping along the tangent lines in increments $\Delta x = h$ according to the formula

$$x_{n+1} = x_n + h \quad \text{and} \quad y_{n+1} = \underline{\hspace{3cm}} \ .$$

36.   In the improved Euler method you use the formulas

$$x_{n+1} = x_n + h,$$
$$z_{n+1} = y_n + hf(x_n,y_n), \quad \text{and}$$
$$y_{n+1} = \underline{\hspace{4cm}} .$$

The formula schemes outlined in Problems 35 and 36 are easily implemented on a computer.

---

34. $-3 + \sqrt{3}i$, $-3 - \sqrt{3}i$, $e^{-3t}[C_1 \cos \sqrt{3}t + C_2 \sin \sqrt{3}t]$, $C \sin \phi$, $\sqrt{3}$,   2,   undercritical, $\frac{2\pi}{\sqrt{3}}$

35. $y_n + hf(x_n,y_n)$ 36. $y_n + \frac{h}{2}[f(x_n,y_n) + f(x_n,z_{n+1})]$

CHAPTER 16 SELF-TEST

1. Show that the function $f(x) = 4 + 2x + x^2 e^x$ is a solution to the differential equation

$$\frac{d^2 y}{dx^2} - 2\frac{dy}{dx} + y = 2e^x + 2x \ .$$

In Problems 2-5, solve the given differential equation.

2. $[x \cos^2(\frac{y}{x}) - y]\ dx + x\ dy = 0$

3. $\frac{dy}{dx} = x^3 e^x + \frac{2y}{x} - 1, \quad x > 0$

4. $e^x(y - 1)\ dx + 2(e^x + 4)\ dy = 0$

5. $y(y^3 - x)\ dx + x(y^3 + x)\ dy = 0$

6. Find the family of orthogonal trajectories to the family of curves given by

$$e^x + e^{-y} = C \ .$$

7. Solve the second order equation

$$x\frac{d^2 y}{dx^2} + \frac{dy}{dx} + x = 0$$

In Problems 8-11, solve the given differential equation.

8. $2y'' - y' - 3y = 0$

9. $9y'' - 12y' + 4y = 0$

10. $y'' + 4y' + 9y = 0$

11. $\frac{d^2 y}{dx^2} + y = \sec^3 x$

12. Find a particular solution of the equation
$$y'' - 9y = x + 2e^{3x} \ .$$

13. A spring is stretched 4 inches by a 2-lb weight. After the weight has come to rest in its new equilibrium position, it is struck a sharp blow that starts it downward at a velocity of 4 ft/sec. If air resistance furnishes a retarding force of magnitude 0.02 of the velocity, find its subsequent motion.

SOLUTIONS TO CHAPTER 16 SELF-TEST

1. $f(x) = 4 + 2x + x^2 e^x$, $f'(x) = 2 + 2xe^x + x^2 e^x$, and

    $f''(x) = 2e^x + 4xe^x + x^2 e^x$. Thus,

    $$f'' - 2f' + f = (2 + 4x + x^2)e^x - (4 + 4xe^x + 2x^2 e^x)$$
    $$+ (4 + 2x + x^2 e^x) = 2e^x + 2x .$$

    Thus, $y = f(x)$ satisfies the differential equation and is, by definition, a solution.

2. From the given equation, we have

    $$\frac{dy}{dx} = \frac{y}{x} - \cos^2\left(\frac{y}{x}\right) = F\left(\frac{y}{x}\right), \quad \text{where} \quad F(v) = v - \cos^2 v \quad \text{and} \quad v = \frac{y}{x} .$$

    Thus, $\frac{dx}{x} + \frac{dv}{v - F(v)} = 0$ becomes $\frac{dx}{x} + \frac{dv}{\cos^2 v} = 0$. The solution of this is

    $$\ln |x| + \tan v = \ln C ,$$

    or, in terms of x and y,

    $$\tan\left(\frac{y}{x}\right) = \ln \frac{C}{|x|} .$$

3. $y' - \frac{2}{x}y = x^3 e^x - 1$, $x > 0$ is linear. An integrating factor is

    $$\rho = e^{\int -\frac{2}{x}\,dx} = e^{-2 \ln x} = x^{-2}, \quad x > 0 .$$

    Thus,

    $$x^{-2}y = \int x^{-2}(x^3 e^x - 1)\,dx$$
    $$= \int xe^{-x}\,dx - \int x^{-2}\,dx$$
    $$= -(x + 1)e^{-x} + x^{-1} + C$$

    or

    $$y = Cx^2 + x - x^2(x + 1)e^{-x} .$$

4. The variables are separable, and the differential equation can be written as

    $$\frac{e^x}{e^x + 4}\,dx + \frac{2}{y - 1}\,dy = 0 .$$

    Integration gives,

    $$\ln(e^x + 4) + 2\ln|y - 1| = \ln C, \quad \text{or} \quad (y - 1)^2(e^x + 4) = C .$$

5. $\frac{\partial}{\partial y}[y(y^3 - x)] = 4y^3 - x$ and $\frac{\partial}{\partial x}[x(y^3 + x)] = y^3 + 2x$,

so the equation is not exact. However, the equation can be rewritten as

$$x\left(\frac{x\ dy\ -\ y\ dx}{y^2}\right) + y(x\ dy + y\ dx) = 0 \ ,$$

or

$$\left(-\frac{x}{y}\right)\ d\left(\frac{x}{y}\right) + d(xy) = 0 \ ,$$

or

$$-u\ du + d(xy) = 0 \ , \quad \text{where} \quad u = \frac{x}{y} \ .$$

Integration gives,

$$-\frac{1}{2}\left(\frac{x}{y}\right)^2 + xy = C \ , \quad \text{or} \quad 2xy^3 - x^2 = 2Cy^2 \ .$$

6. The given family of curves satisfies the differential equation

$$e^x\ dx - e^{-y}\ dy = 0 \ .$$

The family of orthogonal trajectories satisfies the differential equation

$$e^{-y}\ dx + e^x\ dy = 0 \ , \quad \text{or} \quad e^{-x}\ dx + e^y\ dy = 0 \ ,$$

whose solution is

$$e^y - e^{-x} = C_1 \ ,$$

the family of orthogonal trajectories.

7. The dependent variable is missing. We substitute $p = \frac{dy}{dx}$ and

$\frac{dp}{dx} = \frac{d^2y}{dx^2}$, and obtain $x\ \frac{dp}{dx} + p + x = 0$, or $\frac{dp}{dx} + \frac{1}{x}p = -1$. This

last equation is linear in p, and an integrating factor is

$\rho = e^{\int dx/x} = x$. Thus

$$px = \int -x\ dx + C_1, \quad \text{or} \quad x\ \frac{dy}{dx} = -\frac{1}{2}x^2 + C_1.$$

The variables are separable, and we find

$$dy = \left(-\frac{1}{2}x + \frac{1}{x}C_1\right)\ dx \quad \text{gives} \quad y = -\frac{1}{4}x^2 + C_1\ \ln\ |x| + C_2 \ .$$

8.   The characteristic equation is

$$2r^2 - r - 3 = 0$$

$$(2r - 3)(r + 1) = 0$$

or $r = -1, \frac{3}{2}$.   The general solution is

$$y = C_1 e^{-x} + C_2 e^{3x/2} \ .$$

9. $9r^2 - 12r + 4 = 0$

$$(3r - 2)^2 = 0$$

or $r = \frac{2}{3}, \frac{2}{3}$.   The general solution is

$$y = C_1 e^{2x/3} + C_2 x e^{2x/3} \ .$$

10. $r^2 + 4r + 9 = 0$
The roots to the characteristic equation are

$$r = \frac{-4 \pm \sqrt{16 - 36}}{2} = -2 \pm \sqrt{5}\,i \ .$$

The general solution is

$$y = C_1 e^{-2x} \cos \sqrt{5}\,x + C_2 e^{-2x} \sin \sqrt{5}\,x \ .$$

11.   The associated homogeneous equation has the solutions

$$u_1(x) = \cos x \quad \text{and} \quad u_2(x) = \sin x \ .$$

We then have

$$v_1' \cos x + v_2' \sin x = 0$$

$$-v_1' \sin x + v_2' \cos x = \sec^3 x \ .$$

By Cramer's rule:

$$v_1' = \frac{\begin{vmatrix} 0 & \sin x \\ \sec^3 x & \cos x \end{vmatrix}}{\begin{vmatrix} \cos x & \sin x \\ -\sin x & \cos x \end{vmatrix}} = -\frac{\sin x}{\cos^3 x} \ ,$$

$$v_2' = \frac{\begin{vmatrix} \cos x & 0 \\ -\sin x & \sec^3 x \end{vmatrix}}{\begin{vmatrix} \cos x & \sin x \\ -\sin x & \cos x \end{vmatrix}} = \sec^2 x \ .$$

Hence,

$$v_1 = \int \frac{-\sin x\ dx}{\cos^3 x} = -\frac{1}{2}\cos^{-2}x + C_1 = -\frac{1}{2}\sec^2 x + C_1\ ,$$

$$v_2 = \int \sec^2 x\ dx = \tan x + C_2\ ,$$

and

$$y = v_1 u_1 + v_2 u_2$$

$$= -\frac{1}{2}\sec x + C_1 \cos x + \sin^2 x \sec x + C_2 \sin x$$

$$= \sec x\ (\sin^2 x - \tfrac{1}{2}) + C_1 \cos x + C_2 \sin x$$

$$= \sec x\ (\tfrac{1}{2} - \cos^2 x) + C_1 \cos x + C_2 \sin x$$

$$= \tfrac{1}{2}\sec x + C_3 \cos x + C_2 \sin x\ .$$

12. The general solution to the homogeneous equation $y'' - 9y = 0$ is

$$y_h = C_1 e^{-3x} + C_2 e^{3x}\ .$$

Because e^{3x} appears as a term in y_h, we will need a term xe^{3x} in our trial solution. Thus we try

$$y_p = Axe^{3x} + Bx + C\ .$$

The first and second derivatives of y_p are

$$y_p' = 3Axe^{3x} + Ae^{3x} + B$$

$$y_p'' = 9Axe^{3x} + 6Ae^{3x}\ .$$

The result of substituting y_p and its derivatives into the nonhomogeneous equation

$$y'' - 9y = x + 2e^{3x}$$

is

$$\left(9Axe^{3x} + 6Ae^{3x}\right) - 9\left(Axe^{3x} + Bx + C\right) = x + 2e^{3x}$$

$$6Ae^{3x} - 9Bx - 9C = x + 2e^{3x}\ .$$

The last equation is an identity in x if

$$6A = 2,\quad -9B = 1,\quad -9C = 0\ .$$

Thus, $A = \frac{1}{3}$, $B = -\frac{1}{9}$ and $C = 0$.

Therefore,

$$y_p = \tfrac{1}{3}xe^{3x} - \tfrac{1}{9}x$$

is a particular solution.

13. A differential equation of the motion is

$$m\frac{d^2x}{dt^2} = -kx - 0.02\,\frac{dx}{dt} \, ,$$

where $\tfrac{1}{3}k = 2$ or $k = 6$ (by Hooke's law), and $m = \tfrac{2}{32}$. Thus,

$$\frac{d^2x}{dt^2} + \frac{8}{25}\,\frac{dx}{dt} + 96x = 0 \, ,$$

is the differential equation of motion, and the general solution is (see page 1037 of the Finney/Thomas text for undercritical damping)

$$x = Ce^{-0.16t}\,\sin\,(9.8t + \phi) \, .$$

The derivative of the motion is

$$\frac{dx}{dt} = -0.16Ce^{-0.16t}\,\sin\,(9.8t + \phi) + 9.8Ce^{-0.16t}\,\cos\,(9.8t + \phi) \, .$$

From the initial conditions $t = 0$, $x = 0$, $\frac{dx}{dt} = 4$,   we find

$$0 = C\,\sin\,\phi \quad \text{and} \quad 9.8C\,\cos\,\phi = 4 \, .$$

Thus, $\phi = 0$ and $C \approx 0.41$.   Hence,

$$x = 0.41e^{-0.16t}\,\sin\,(9.8t)$$

describes the motion. The motion is oscillatory with undercritical damping.